GRASSLANDS OF THE UNITED STATES

Their Economic and Ecologic Importance

Grasslands

A SYMPOSIUM OF THE AMERICAN

THE IOWA STATE

of the United States

Their Economic and Ecologic Importance

FORAGE AND GRASSLAND COUNCIL

Edited by   Howard B. Sprague

UNIVERSITY PRESS, AMES, IOWA, U.S.A.

1 9 7 4

© 1974 The Iowa State University Press
Ames, Iowa 50010. All rights reserved

Composed and printed by The Iowa State University Press

First edition, 1974

Library of Congress Cataloging in Publication Data

Sprague, Howard Bennett, 1898–
   Grasslands of the United States: their economic and ecologic importance.

   Includes bibliographies.
   1. Grasslands—United States—Congresses.   2. Forage—Congresses.
3. Forage plants—United States—Congresses.   I. American Forage and
Grassland Council.   II. Title.
SB193.S68   333.7′4′0973   74–620
ISBN 0-8138-0745-X

N. D. Bayley
Director, Science and Education
Agricultural Research Service
U.S. Department of Agriculture

B. D. Blakely
Chief Agronomist
Soil Conservation Service
U.S. Department of Agriculture

Donald F. Burzlaff
Chairman, Department of Range and
Wildlife Sciences
Texas Tech University

Lawrence V. Compton
Biologist
Soil Conservation Service
U.S. Department of Agriculture

Olan W. Dillon, Jr.
Biologist
Soil Conservation Service
U.S. Department of Agriculture

S. H. Dobson
Extension Professor, Department of
Agronomy
North Carolina State University

L. M. Glymph
Assistant Director
Soil and Water Conservation Re-
search Division
Agricultural Research Service
U.S. Department of Agriculture

Chester H. Gordon
Leader, Nutrition Investigation
Dairy Cattle Research Branch
Agricultural Research Service
U.S. Department of Agriculture

Wade H. Hamor
Biologist
Soil Conservation Service
U.S. Department of Agriculture

Maurice E. Heath
Associate Professor, Department of
Agronomy
Purdue University

H. J. Hodgson
Principal Agronomist
Cooperative State Research Service
U.S. Department of Agriculture

R. E. Hodgson
Director, Animal Research Branch
Agricultural Research Service
U.S. Department of Agriculture

H. N. Holtan
Director, Hydrograph Laboratory
Agricultural Research Service
U.S. Department of Agriculture

George D. Lea
Chief, Division of Range
Bureau of Land Management
U.S. Department of the Interior

David A. Mays
Forage Specialist, Soils and Research Branch
Tennessee Valley Authority

L. A. Moore
Former Director, Dairy Husbandry Research
Agricultural Research Service
U.S. Department of Agriculture

Lloyd E. Partain
Assistant to the Administrator for Environmental Development
Soil Conservation Service
U.S. Department of Agriculture

Donald A. Price
Director, U.S. Sheep Experiment Station
Sheep and Fur Animal Research Branch
Animal Science Research Division
Agricultural Research Service
U.S. Department of Agriculture

P. A. Putnam
Chief, Beef Cattle Research Branch
Agricultural Research Service
U.S. Department of Agriculture

Robert S. Rummell
Staff Assistant, Division of Range Management
Forest Service
U.S. Department of Agriculture

Howard B. Sprague
Agricultural Consultant
Former Head, Department of Agronomy
The Pennsylvania State University

C. H. Wadleigh
Science Adviser, Office of the Administrator
Agricultural Research Service
U.S. Department of Agriculture

John B. Washko
Professor, Department of Agronomy
The Pennsylvania State University

R. E. Williams
Chief Range Conservationist
Soil Conservation Service
U.S. Department of Agriculture

CONTENTS

vii

FOREWORD

THE VALUE of forage production in the United States is conservatively estimated at $12 billion. In addition, forages add beauty to the landscape, aid in preserving environmental quality, effect erosion control, and offer immeasurable recreation potential.

By 1980 the United States will need at least 30 percent more meat and milk to feed the 25 million more people expected by that time. Beef cattle rank among the top four income-producing commodities in 41 states. Dairy products rank first in 13 states and among the top four in 39 states. Thus few feed resources are as important to the welfare of our nation as forages. Nevertheless, this area of crop production has been given a relatively low priority in the scheme of things. To meet the challenge immediately before us, the efficiency and use of this resource must be substantially improved. Currently, we are only realizing approximately 20 percent of our forage and grassland potential; thus this book is most timely and valuable.

The American Forage and Grassland Council is pleased to be the sponsor of this publication. It is my pleasure and privilege to commend and thank the editor and authors for their contributions.

J. RITCHIE COWAN
PRESIDENT
AMERICAN FORAGE AND GRASSLAND COUNCIL

THIS VOLUME has been prepared under the sponsorship of the American Forage and Grassland Council. The Council is dedicated to promoting the effective use of all types of forages; to providing a forum for all workers from public and private sectors with an interest in forage; to the collection and analysis of information on forage resources; and to the encouragement of industry, research, and educational organizations to contribute to efficient production, marketing, and utilization as may be in the best interests of the agricultural economy and the best conservation of available natural resources. The Council is composed of individual members, corporate members, professional societies, and institutions.

The Council is deeply concerned with developing a broad understanding of the role filled by grasslands of all types in the total economy and with the conservation and effective management of these grassland resources to protect and develop our environmental heritage. This volume was compiled to provide a comprehensive inventory and evaluation of the grasslands in the forty-eight conterminous states, with the hope that similar studies may be made of grassland resources and values in other regions of the continent.

The grasslands of the United States as a whole are immense and highly diverse. To deal with such vast resources embracing all the geographic regions of the nation, selected specialists were enlisted to present the essential character of the many sectors and to outline the significance of each in the national economy. In addition, the general significance of grasslands, both direct and indirect, is dealt with in several chapters. The authors of this volume, representing many disciplines and areas of public responsibility, are national authorities on the subjects covered by their contributions. This symposium volume brings together a wealth of information on grasslands that would otherwise be difficult to assemble and interpret. The illustrations provide representative views of our grasslands and their uses.

Special recognition is due Kevin G. Hayes of The Pennsylvania State University for his assistance in the preparation of this volume.

HOWARD B. SPRAGUE

GRASSLANDS OF THE UNITED STATES

Their Economic and Ecologic Importance

1

SIGNIFICANCE OF OUR GRASSLANDS

HOWARD B. SPRAGUE

THE IMPORTANCE of grasslands in our national life has been difficult to assess. As a result, these lands have generally been accorded far
less consideration than warranted by the important roles they fulfill
throughout the nation. Judgments of knowledgeable specialists on
prudent conservation and management of the exceedingly varied
categories of these land areas are fragmented and dispersed.

Several reasons account for the lack of essential information per-
taining to grasslands. Unlike other agricultural production data,
neither the U.S. agricultural census nor the federal-state Crop Reporting Service provides any direct data on the productivity of grasslands
except for relatively small areas harvested by man for hay and grass
silage. Harvested crops and livestock production are enumerated by
the Census Bureau and the Crop Reporting Service, but grassland
production harvested by livestock does not come within the scope of
authorized agricultural production reporting. The problem is further
complicated by the relatively large expanses of grazing lands that fall
within the public domain, federal and state. The majority of
federal grazing lands, largely administered by the U.S. Forest Service
and Bureau of Land Management, are used by private ranchers under
leasing arrangements or permits and their contribution to livestock
production is very significant. Nevertheless, information on contributions of such resources to our economy is not easily located nor correlated with the use and productivity of privately owned grasslands.

To comprehend the significance of our grasslands, two dimensions
of this resource should be examined, the total acreage of such lands
and their geographic distribution. The different broad classes of
U.S. grasslands according to use and acreage are shown in Figure 1.1.

CHAPTER 1

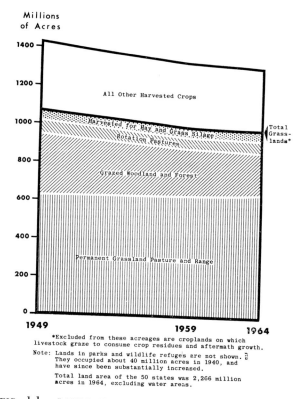

FIG. 1.1. LANDS OF THE UNITED STATES USED FOR GRASSLANDS AND HARVESTED CROPS.

In evaluating this chart, remember that all grasslands are subject to multiple uses irrespective of ownership and geographic location. In addition to their economic production the grasslands are effective for soil and water conservation on the lands themselves, they contribute to regional water conservation, they support wildlife and game, and they constitute an important part of the total environment for the esthetic pleasure of man. Grasslands occupy about one-half the total land area of the 48 contiguous states. They warrant the attention and recognition of all concerned with our national resources.

Figure 1.2 indicates the geographic distribution of privately owned grasslands. The distribution of publicly owned grasslands, which account for about one-third of the national total, occurs very largely in the western states west of the 100° meridian. Collectively, the privately and publicly owned grasslands constitute a major feature

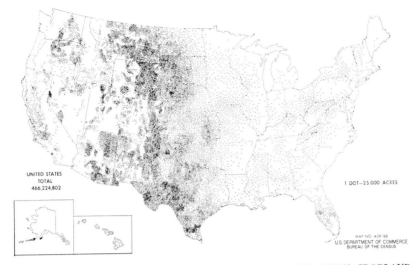

FIG. 1.2. PRIVATELY OWNED PASTURE OTHER THAN CROPLAND AND WOODLAND.

of the national landscape in all sections of the nation. Therefore grasslands are of concern directly or indirectly to nearly everyone whether urban, suburban, or rural. These resources are part of the national heritage and should be wisely managed.

A reasonably adequate evaluation of the significance of the national grasslands should include the following aspects:

1. The support they provide to the national livestock industry.
2. Their use in soil and water conservation.
3. Their role in regional water resource development and protection in service to people of nearby areas.
4. Their importance in balanced programs of wildlife and game conservation and protection.
5. Their contribution to outdoor recreation and pleasure.
6. Alternative uses of the various types of grasslands, i.e., the degree to which present occupation of land by grass cover and its associated use constitutes the most effective ecological adjustment of these land types for balanced and prudent management of the total environment.

The successive chapters of this volume deal with these important aspects of grasslands in our national life.

2

OUR GRAZING LAND RESOURCES

B . D . B L A K E L Y *and* R . E . W I L L I A M S

PRIVATELY OWNED grazing land (range, pasture, and grazed forests) constitutes about 43 percent of the total land included in the 1967 USDA conservation needs inventory (4). The major part of forage required for the nation's livestock comes from this land.

The need for meat and dairy products in the year 2000 is expected to increase over that of 1960 by almost 31.5 billion pounds of beef, 48 million pounds of mutton, 11.8 billion pounds of pork, and 58 billion pounds of milk. Our growing need for meat and dairy products, the increasing problem of pollution associated with concentrating large numbers of animals in feedlots, and the growing numbers of livestock (including horses on recreation ranches and farms) all point to a need for more grazing land, not less. This need must be met by improving and properly managing the grazing land we have.

About 698 million acres of the nation's grazing land is privately owned. An additional 262.7 million acres is federally owned, managed by the Forest Service and the Bureau of Land Management. The number of acres of different grazing lands are (4):

Privately owned:	*Acres*
Rangeland	380,521,400
Pastureland	103,123,900
	(3.2 million acres irrigated)
Cropland used for hay	77,674,600
or pasture	(13 million acres irrigated)
Grazed forests	137,241,100
Federally owned:	
Rangeland	262,742,000(2)

Table 2.1 shows 20 major land resource areas in the United States and the different kinds of grazing land in each (3).

6

TABLE 2.1. *Land resource regions of the conterminous United States*

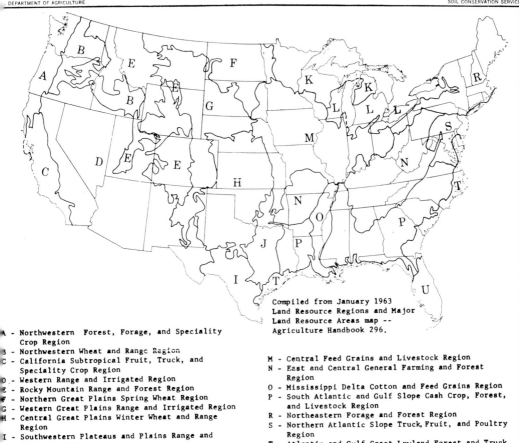

DEPARTMENT OF AGRICULTURE

SOIL CONSERVATION SERVICE

Compiled from January 1963
Land Resource Regions and Major
Land Resource Areas map --
Agriculture Handbook 296.

A - Northwestern Forest, Forage, and Speciality
  Crop Region
B - Northwestern Wheat and Range Region
C - California Subtropical Fruit, Truck, and
  Speciality Crop Region
D - Western Range and Irrigated Region
E - Rocky Mountain Range and Forest Region
F - Northern Great Plains Spring Wheat Region
G - Western Great Plains Range and Irrigated Region
H - Central Great Plains Winter Wheat and Range
  Region
I - Southwestern Plateaus and Plains Range and
  Cotton Region
J - Southwestern Prairies Cotton and Forage Region
K - Northern Lake States Forest and Forage Region
L - Lake States Fruit, Truck, and Dairy Region

M - Central Feed Grains and Livestock Region
N - East and Central General Farming and Forest
  Region
O - Mississippi Delta Cotton and Feed Grains Region
P - South Atlantic and Gulf Slope Cash Crop, Forest,
  and Livestock Region
R - Northeastern Forage and Forest Region
S - Northern Atlantic Slope Truck, Fruit, and Poultry
  Region
T - Atlantic and Gulf Coast Lowland Forest and Truck
  Crop Region
U - Florida Subtropical Fruit, Truck Crop, and
  Range Region

Acres of Privately Owned Land Available for Grazing
Thousand Acre Units

Land Resource Region	Total Land In Inventory	Pastureland	Rangeland	Grazed Forest	Cropland Used For Hay or Pasture	Percent of Total Acres
A	25656.9	783.9	708.9	2507.5	1546.0	22
B	37969.7	1223.3	15273.6	4467.1	2553.0	54
C	30237.2	1269.0	7816.7	5473.8	1471.2	53
D	141872.9	1888.1	95130.6	21555.2	3412.1	85
E	59348.2	2241.1	25731.9	13929.8	3821.2	77
F	76625.0	2484.6	25379.0	569.4	6380.3	45
G	121933.0	1657.8	89898.2	6331.4	4220.9	83
H	140875.9	2061.7	67679.2	3372.8	3135.0	53
I	42051.9	805.7	32000.3	4280.9	54.7	88
J	36618.5	9411.8	8607.5	7211.9	777.7	72
K	54792.4	2925.5	11.0	2675.5	4498.0	18
L	33205.6	1898.8	-	524.9	4847.6	23
M	172387.3	21211.5	5426.7	9041.0	18728.5	31
N	127903.1	21044.0	1605.4	16134.3	9819.3	38
O	19838.0	1207.4	25.3	1468.0	189.4	14
P	159429.2	18657.3	81.2	25886.3	6288.4	32
R	67767.3	3694.8	-	1322.1	5213.6	15
S	24753.8	2339.9	-	524.5	2129.6	20
T	41088.3	2276.2	2643.3	3701.0	1945.3	26
U	17572.6	1979.2	1909.8	5360.0	53.4	52
Total	1,437,985.0	101,740.0	380,139.0	136,790.0	77,646.0	48
Alaska	2424.7	.8	208.4	.4	16.2	9
Hawaii	3640.3	677.4	-	450.7	.6	31
Carib.Area	2026.3	705.7	174.0	-	11.8	26

In 1967 eight agencies in the USDA cooperated with the Indian Service in the Department of Interior to conduct a conservation needs inventory on nonfederal and Indian lands (4). Included were kinds and amounts of treatment needed and kinds of soil on pastureland, rangeland, and grazed forests.

The inventory was made on a 2 percent sample. In much of the country, 160-acre sample areas were used; in the irrigated areas of the West, 40-acre sample areas were used. On the basis of the soils and prevailing farming enterprise, a soil conservationist determined the land use and the conservation treatment most needed to maintain or improve soil and water resources and protect the forage-producing potential. This sample information was expanded to include the total area in farms, ranches, and Indian-owned lands in each county of the nation.

Eight categories for pastureland and rangeland were used to reflect the most needed treatment:

1. *Treatment adequate*—Vegetative cover was in high, fair, or better condition. Adequate treatment for rangeland was based on the potential of the site.
2. *No treatment feasible*—Feasibility was based on a reasonable return on investment.
3. *Change in land use*—The area should be planted to trees.
4. *Protection only*—The desired vegetation, though present, was usually overgrazed. This could be corrected by livestock management and/or installation of watering facilities to improve grazing distribution.
5. *Improvement only*—The grass cover was inadequate but could be improved or restored by fertilization and good management practices.
6. *Improvement and brush control*—Control of woody or noxious plants through mechanical or chemical measures and stand improvement was needed.
7. *Reestablishment of vegetative cover*—Vegetation was in such poor condition that complete reseeding was needed.
8. *Reestablishment and brush control*—In addition to reseeding, brush-control measures were also needed.

PERMANENT PASTURE

IN MOST CASES, permanent pasture produces introduced forage species. It may require management practices such as reseeding, liming, fertilizing, and proper grazing use. If properly managed, permanent pasture could supply a considerable part of livestock forage needs. In addition to supplying forage, a good pasture keeps runoff sediment-free and benefits wildlife.

The 103.1 million acres of pasture is about 15 percent of the total acreage available for grazing. Nearly 75 percent of the permanent pastureland is in land capability classes I, II, III, and IV and will respond to treatment if it is needed by the farmer to maintain or expand his number of livestock (1).

About 46 percent of the pasture can be improved by proper grazing, fertilization, or brush control. Only 33 percent needs reseeding to better grasses or legumes. Table 2.2 summarizes the treatment needs for pasture by major land resource regions (3).

Not every acre of pasture on every farm needs to be managed for maximum production. The number of livestock and available acres of pasture determine the needed level of management. However, the degree of management often reflects the economic return to the livestock producer. Well-managed pasture is basic to a good conservation program on livestock farms in the East. It is also an integral part of the forage program on many ranches in the West. Planning a full season of grazing by using grasses that are most productive at varying times during the grazing season should be emphasized.

CROPLAND FOR HAY OR PASTURE

THE ACREAGE of cropland used for hay or pasture reported in the 1967 inventory amounted to about 77.6 million acres. This is about 11 million less acres than those used during the 1950–60 period. According to the inventory, hay or pasture cropland represents only 11 percent of total forage resources but supplies nearly all the winter forage for beef and sheep and a considerable part of that needed for dairy cattle. In addition, it supplies feed during a part of the year when other kinds of forage are limited (4).

This source of forage is of major importance in land resource regions K, L, M, R, and S (see Table 2.2) inasmuch as it accounts for nearly 50 percent of their total grazing land. Three-fourths of this hay or pastureland occurs on some of the best soils in the country. If properly fertilized, seeded to adapted species, and managed, the carrying capacity can be as much as two animal units per acre. For this reason, this land is a valuable economic source of forage on many livestock farms.

TABLE 2.2. *Conservation treatment needs for pasture by major land resource regions*

Land resource region	Land capability group	Total acres	Adequately treated	No treatment feasible	Change in land use	Protection only	Improvement only	Improvement and brush control	Reestablishment of vegetative cover	Reestablishment and brush control
						(000 acres)				
A	I-IV	427.1	109.0	7.5	2.9	47.8	74.2	34.4	76.1	75.1
	V-VII	356.1	58.1	3.2	89.0	58.1	45.0	32.7	25.7	44.4
B	I-IV	869.3	157.0	3.9	1.8	158.2	420.9	37.0	65.4	25.1
	V-VIII	354.0	81.9	5.5	0.3	98.1	85.7	19.3	22.3	41.0
C	I-IV	1,133.7	406.5	172.6	2.6	280.5	194.1	5.6	66.2	5.6
	V-VIII	135.3	42.8	4.9	...	77.9	6.2	1.8	1.4	0.2
D	I-IV	1,058.0	331.2	18.5	0.4	206.0	340.2	13.8	126.1	21.6
	V-VIII	830.1	190.8	26.5	0.1	310.3	195.2	30.4	57.9	18.9
E	I-IV	1,575.7	531.8	3.7	0.1	266.5	672.6	22.9	61.2	16.7
	V-VIII	665.4	294.5	0.3	0.2	159.2	184.6	9.7	12.2	4.7
F	I-IV	2,146.1	809.3	0.9	...	572.4	503.8	20.0	206.0	33.7
	V-VIII	338.5	151.6	0.8	...	82.3	59.0	3.9	38.7	1.9
G	I-IV	1,248.7	547.5	0.4	...	375.5	222.8	6.5	95.4	0.5
	V-VIII	409.1	144.6	1.9	...	129.8	52.8	14.9	65.2	...
H	I-IV	1,608.1	656.4	3.9	...	290.4	424.4	39.3	174.2	19.5
	V-VIII	453.6	141.1	5.6	...	85.3	85.0	20.3	110.0	6.4
I	I-IV	730.4	359.7	1.2	...	89.4	153.0	21.5	50.9	54.6
	V-VIII	75.3	34.6	0.2	...	15.8	13.9	4.0	4.6	2.3
J	I-IV	7,360.2	1,753.9	11.1	0.2	744.2	2,349.7	390.5	1,228.6	882.0
	V-VIII	2,051.6	392.6	8.4	...	176.1	636.1	176.1	269.4	392.9
K	I-IV	2,327.2	521.9	27.1	16.8	398.3	345.7	273.2	512.0	232.1
	V-VIII	598.2	91.1	8.7	81.0	137.8	73.3	58.8	103.2	44.4
L	I-IV	1,579.7	439.5	28.2	34.8	192.0	332.5	99.4	276.9	176.4
	V-VIII	319.1	59.9	10.6	69.9	42.9	48.9	8.7	47.0	31.0
M	I-IV	16,500.9	4,832.7	158.9	39.0	2,375.8	4,256.2	764.1	2,647.2	1,427.0
	V-VIII	4,710.6	858.3	139.6	188.7	607.9	1,039.7	316.5	825.9	733.9
N	I-IV	13,350.1	4,466.0	39.9	62.2	1,141.1	3,708.6	757.3	1,970.0	1,205.0
	V-VIII	7,693.9	1,414.9	112.1	996.1	562.0	1,994.9	699.3	880.0	1,034.6
O	I-IV	1,086.9	332.1	2.8	...	113.4	257.9	26.3	296.4	58.0
	V-VIII	120.5	40.5	...	...	15.3	27.2	9.6	17.4	10.5
P	I-IV	15,416.9	4,229.6	25.4	21.5	1,848.9	5,162.5	725.9	2,244.8	1,158.3
	V-VIII	3,240.4	534.0	8.9	172.3	281.2	1,158.8	149.7	504.4	431.1
R	I-IV	2,683.0	879.5	144.7	119.1	139.0	506.1	218.4	319.3	356.9
	V-VIII	1,011.9	128.1	214.3	266.6	40.7	105.1	80.6	54.7	121.7
S	I-IV	1,671.6	587.2	12.1	9.3	135.2	600.3	74.9	171.3	81.2
	V-VIII	668.3	132.0	17.1	59.3	58.9	249.8	41.8	72.4	37.3
T	I-IV	2,171.0	646.7	4.3	...	429.3	792.4	38.1	208.4	51.8
	V-VIII	105.2	42.4	0.6	...	21.7	32.3	0.2	4.5	3.4
U	I-IV	1,814.6	627.1	2.1	0.2	671.1	356.5	78.3	51.4	27.9
	V-VIII	164.6	66.7	8.3	...	45.7	36.6	5.1	1.1	1.2
Total	I-IV	76,759.2	23,224.6	669.2	310.9	10,475.0	21,674.4	3,647.4	10,847.8	5,909.0
	V-VIII	24,301.7	4,900.5	577.5	1,923.5	3,007.0	6,130.1	1,683.4	3,118.0	2,961.8
Alaska	I-IV	.8	...	...	...	0.4	...	...	...	...
	V-VIII	none								
Hawaii	I-IV	237.7	48.6	0.9	...	0.1	57.5	40.9	54.1	35.6
	V-VIII	439.7	61.4	16.0	2.2	6.5	54.8	91.3	161.3	46.3
Caribbean area	I-IV	208.6	85.2	0.2	0.5	17.6	25.8	17.7	30.7	31.0
	V-VII	497.1	161.8	4.9	3.6	33.0	64.1	48.9	85.8	95.9

Source: USDA (3).

GRAZED FORESTS

ACCORDING to the conservation needs inventory, chaparral areas and areas with at least 10 percent in trees capable of producing timber or other wood products are classified as forests (4).

There are more than 137 million acres of grazed forests, of which nearly 29 million are considered adequately treated for grazing purposes; about 64 million acres (47 percent) could be improved for grazing. The remaining 43 million acres should be protected from grazing. Most of the latter is in land resource regions K, L, M, N, R, and S. About 70 percent of the grazed forests in land resource regions D, H, I, and J could be treated to improve grazing capacity.

The capacity of these forests for producing forage varies from very little under thick stands of trees to a considerable amount if the tree stand is near the 10 percent level. About 47 million acres of the 64 million acres of noncommercial forests are grazed.

RANGELAND

RANGELAND is dominated by native vegetation (grasses, grasslike plants, forbs, and shrubs) that is suitable for grazing. This includes natural grasslands, savannas, many wetlands, tundra, and certain forb and shrub communities. Some rangelands have been or may be seeded to long-lived perennial native or introduced species.

While pastureland is a community of one or more introduced forage species, rangeland is a community of several to many kinds of native plants. Native plant communities are long-lived and persist and improve under judicious grazing management.

The diversity of plants on rangeland and related grazing lands makes it valuable for several compatible secondary uses, e.g., a habitat for big-game animals and other wildlife. Good grazing land management is good wildlife management for many important kinds of animals.

Range conditions and conservation treatment are adequate on 111.9 million acres (29 percent) of the 380.5 million acres of rangeland reported in the conservation needs inventory. No treatment is feasible on 16.9 million acres, or 5 percent, because of extremely adverse soil or climatic conditions. The following treatments are needed on 251.2 million acres (66 percent): 123.3 million acres (32 percent) need protection from overgrazing, 91.5 million acres (25 percent) need improvement in plant cover, including 51.5 million acres of brush control, and 36.4 million acres (9 percent) need complete reestablishment of plant cover, including 24.6 million acres of brush control.

Table 2.3 summarizes the treatment needed by major land resource regions to protect soil and water resources of the rangelands and to restore them to their potential production (3). Rangeland treatment sometimes can be combined with improvement of wildlife habitats. For example, brush of value for cover and food for deer can be left along rough ridges and watercourses. Large level areas

TABLE 2.3. *Conservation treatment needs for range by major land resource regions*

Land resource region	Land capability group	Total acres	Adequately treated	No treatment feasible	Change in land use	Protection only	Improvement only	Improvement and brush control	Reestablishment of vegetative cover	Reestablishment and brush control
						(000 acres)				
A	I-IV	216.1	90.7	...	...	53.1	8.6	5.8	45.0	12.7
	V-VIII	492.9	49.2	116.3	...	268.7	7.8	10.0	27.9	13.0
B	I-IV	2,813.1	680.1	23.3	0.7	534.9	212.1	489.5	419.7	452.8
	V-VIII	12,460.5	3,108.2	418.2	2.1	3,677.1	1,890.4	2,032.9	554.9	776.7
C	I-IV	2,775.2	863.6	75.1	6.5	1,154.5	406.9	68.5	74.9	125.2
	V-VIII	5,041.4	988.4	1,635.9	4.3	1,680.2	414.5	196.4	53.8	68.0
D	I-IV	2,656.3	389.2	71.4	5.5	802.5	195.6	648.5	181.2	362.4
	V-VIII	92,474.3	13,941.6	9,692.5	39.8	32,810.3	14,560.1	12,430.3	2,873.6	6,126.1
E	I-IV	4,802.8	1,756.0	6.3	0.6	1,907.8	283.9	579.4	180.0	88.7
	V-VIII	20,929.1	7,353.9	693.9	1.0	8,387.4	1,090.5	2,568.3	254.2	579.7
F	I-IV	12,971.9	5,498.6	6.3	...	5,106.3	1,893.9	292.9	166.9	7.1
	V-VIII	12,407.0	5,916.4	281.6	...	5,101.1	888.3	155.2	60.4	4.0
G	I-IV	23,123.6	8,580.9	28.1	...	9,549.5	2,474.0	1,534.8	869.3	87.0
	V-VIII	66,774.6	25,268.2	2,165.9	...	26,794.2	6,797.5	4,016.1	1,359.8	372.9
H	I-IV	27,071.2	10,121.1	150.5	0.4	7,085.1	1,383.5	5,624.5	1,297.7	1,408.3
	V-VIII	40,608.0	15,113.7	1,097.7	2.1	9,538.7	2,510.9	9,547.3	1,020.3	1,777.3
I	I-IV	12,114.2	1,646.7	19.2	...	585.5	415.4	2,739.3	313.0	6,394.9
	V-VIII	19,886.1	5,869.2	192.5	0.2	3,002.2	1,717.6	4,376.0	303.2	4,425.2
J	I-IV	4,503.2	726.8	14.2	...	968.2	668.2	803.4	696.5	625.5
	V-VIII	4,104.3	862.8	20.5	...	1,328.4	579.3	744.4	344.5	224.5
K	I-IV	6.7	2.8	...	...	0.5	1.4	1.8	trace	trace
	V-VIII	4.3	0.5	...	...	0.3	0.1	3.3	trace	trace
L	none									
M	I-IV	3,179.1	1,004.2	8.4	0.2	1,096.4	318.9	416.8	223.2	110.9
	V-VIII	2,247.6	548.8	27.4	1.3	754.8	240.9	492.4	95.7	86.5
N	I-IV	994.3	157.9	0.2	0.5	86.4	204.2	273.3	153.0	118.7
	V-VIII	611.0	110.0	4.5	0.2	48.8	71.0	171.1	96.6	108.9
O	I-IV	11.3	1.8	...	...	...	4.9	4.7	...	...
	V-VIII	14.0	1.7	...	...	...	...	12.3	...	
P	I-IV	49.6	0.6	...	...	8.7	12.1	22.0	5.9	0.3
	V-VIII	31.6	14.7	...	...	...	...	12.3	...	...
R	none									
S	none									
T	I-IV	1,536.6	302.7	1.8	...	288.3	227.0	512.8	59.9	144.0
	V-VIII	1,106.7	541.4	36.5	...	250.7	104.6	116.0	17.8	39.7
U	I-IV	1,366.6	203.3	39.0	...	355.9	193.9	490.2	13.3	71.0
	V-VIII	543.2	45.4	129.4	...	81.6	208.1	59.5	2.6	16.5
Total	I-IV	100,191.8	32,027.0	443.8	14.4	29,583.6	8,904.5	14,508.2	4,699.5	10,009.5
	V-VIII	279,736.6	79,734.1	16,512.8	51.0	93,727.4	31,087.2	36,939.0	7,066.1	14,619.0
Alaska	I-IV	53.5	33.4	...	...	7.0	...	13.0	...	...
	V-VIII	144.9	108.8	...	...	15.8	...	30.3	...	...
Hawaii	none									
Caribbean area	I-IV	43.6	18.9	0.2	...	3.7	5.1	8.6	4.3	2.8
	V-VIII	130.5	43.1	2.2	0.2	7.3	10.0	26.4	18.3	23.0

Source: USDA (3).

can be cleared of brush in strips, improving forage for livestock yet leaving some brush for food and cover for wildlife. Areas on which brush is controlled produce more forage for livestock, and the edges between the cleared and uncleared strips favor deer and upland game. Plant cover on native grazing land is significantly related to erosion control and runoff as well as sediment loads in streams and lakes.

Because native grazing land offers open space and diversity in plants, topography, domestic animals, and wildlife, it attracts a growing number of outdoor recreationists. In considering conservation needs and treatment, both primary and compatible secondary uses of the land should be considered.

TABLE 2.4. *Acreage of land grazed, acreage suitable for grazing and percent of suitable acreage needing and susceptible to range improvement practices, by state, for the public land administered by the Forest Service and the Bureau of Land Management, 1966*

State	Forest Service			Bureau of Land Management			Percent of total feed units furnished Forest Service and Bureau of Land Management
	Area grazed	Acreage suitable for grazing	Percent needing range improvement practices	Area grazed	Acreage suitable for grazing	Percent needing range improvement practices	
	(000 acres)			*(000 acres)*			
Montana	7,670	2,740	79	7,790	7,950	49	7
Idaho	11,800	4,120	68	11,600	12,100	71	17
Wyoming	6,830	2,720	75	17,300	17,300	85	16
Colorado	12,400	5,370	80	7,800	7,800	78	6
New Mexico	8,350	4,890	92	13,500	13,600	90	17
Arizona	11,300	7,460	95	12,500	12,500	39	27
Utah	7,120	3,040	83	21,100	21,800	64	28
Nevada	4,860	1,870	81	43,200	44,400	82	49
Washington	4,440	1,160	90	233	285	62	2
Oregon	9,860	6,060	94	13,900	13,900	33	13
California	12,000	3,460	80	8,410	8,830	77	4
11 western states	96,656	42,882	85	157,211	160,408	70	12
Other states	8,784	6,453	95	92	364	20	…
United States	105,440	49,335	86	157,303	160,772	70	3

Treatments defined in the conservation needs inventory were related in some cases to cultural treatments such as seeding and brush control (4). The most vital aspect of grazing management, however, is the proper distribution of grazing animals. Proper grazing use (including kinds and numbers of livestock, season of use, and planned grazing systems) is essential in all treatment needs discussed.

FEDERAL RANGELAND

THE FOREST SERVICE of the USDA administers 105.4 million acres of federal rangeland grazed by domestic livestock. The Bureau of Land Management of the Department of the Interior administers 157.3 million acres of other federal rangeland. Table 2.4 shows, by states, acres grazed on public land administered by the Forest Service and Bureau of Land Management, the percentage needing range improvement practices, and the percentage of total feed furnished by federal rangeland.

Needs for range improvement practices were not determined by the same criteria used in the conservation needs inventory. Indications are that most of the federal rangeland requires conservation treatment just as most of the privately owned rangeland does. Table 2.4 also shows, in percentage of feed units produced, that forage produced on federal rangeland is especially important in several western states. Grazing on federal rangeland is vitally important to many ranchers and communities in the rangeland areas.

REFERENCES

1. Klingebiel, A. A., and P. H. Montgomery. 1961. *Land-Capability Classification*. SCS, USDA Agr. Handbook 210.
2. University of Idaho. 1966. Forage study for the Public Land Law Review Commission.
3. USDA. 1963. Agr. Handbook 296.
4. USDA. 1967. Conservation needs inventory.

3

GRASSLANDS IN RELATION TO
WATER RESOURCES

C. H. WADLEIGH, L. M. GLYMPH, *and*
H. N. HOLTAN

VERDANT FORESTS ensure clear water in the brooks and creeks. Our foresters have done a commendable job of studying and relating the benefits of forest cover to the hydrology of an area, including the protection of water quality. Indications are that grass cover provides similar benefits to water resources, but the effects of grasslands have not been documented as well as those of forests in relation to water.

HYDROLOGIC CYCLE

To ATTAIN a general perspective of the subject, we must first review the operation of the hydrologic cycle and then consider how the various steps and processes in the hydrologic continuum are affected by grass cover as compared with other kinds of cover or no cover.

Ackermann, et al. (2) have presented a descriptive representation of the hydrologic cycle (Fig. 3.1). This figure provides at a glance an indication of the various paths followed by water as it moves through the atmosphere, the hydrosphere, and the lithosphere. Actually, the system is highly complex with no specific beginning or end.

Water, evaporating from the land and its plant cover as well as from lakes, rivers, and oceans, moves into the atmosphere wherein it may be carried great distances. Precipitation brings water to the land where it enters into many diverse paths such as surface runoff, streamflow to lakes or oceans, soil moisture accretion, absorption by plant roots, deep percolation, groundwater recharge, and lateral flow in geologic formations that may emerge as seeps and springs.

Diagrams in this chapter reprinted with permission from *Handbook of Applied Hydrology*, ed. Ven Te Chow. Copyright © 1964 by McGraw-Hill, Inc.

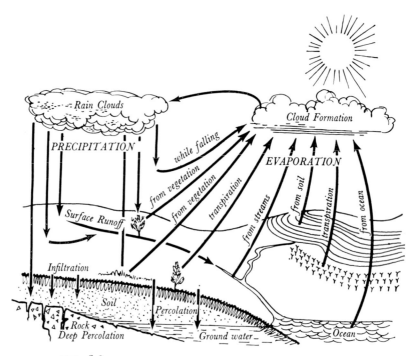

FIG. 3.1. THE HYDROLOGIC CYCLE, A DESCRIPTIVE REPRESEN-
TATION (2).

It is of interest also to glance at Horton's (17) representation of
the water cycle (Fig. 3.2), which indicates how the various processes
involved fit together under four main arcs: atmospheric transport,
precipitation, surface disposition, and evaporation. The nature of
the vegetative cover may have quite an effect on many of these processes.

To gain a quantitative perspective of the magnitude of the water
cycle for the 48 contiguous states, it is further helpful to scan the chart
in Figure 3.3 organized by Wolman (32) from data presented by Acker-
man and Lof (1). It is especially germane to our objectives that 70
percent of the average annual precipitation onto the United States
becomes evapotranspiration from nonirrigated land. Thus the only
way that economic benefit may accrue from the lion's share of our
water supply is through the growth of economic plants. Also, the
value that may be assigned to water consumed in evapotranspiration
will be largely determined by the amount of economic plant product
produced per inch of water evaporated. For example, climatic con-

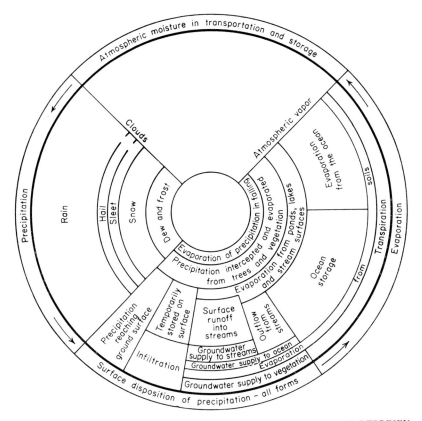

FIG. 3.2. THE HYDROLOGIC CYCLE, A QUALITATIVE REPRESEN-
TATION (17).

ditions in central New York State determine that a 1-acre potato field
in that area will use about 18 inches of water in evapotranspiration
during the growing season regardless of the size of the crop produced.
These 18 acre-inches of water amount to about 500,000 gallons. A
well-fertilized potato field protected from pests in that region will
produce 500 bushels per acre. A neglected field may produce only 50
bushels per acre at the same level of evapotranspiration. In the latter
case, 10,000 gallons of water would be used for each bushel of potatoes
produced. It would be exceedingly difficult to argue that such a situation constituted maximum beneficial use of water. In fact, water consumed in producing such a poor potato yield cannot have a value
above one cent per 1,000 gallons.

Although this reference to a hypothetical case involving potatoes

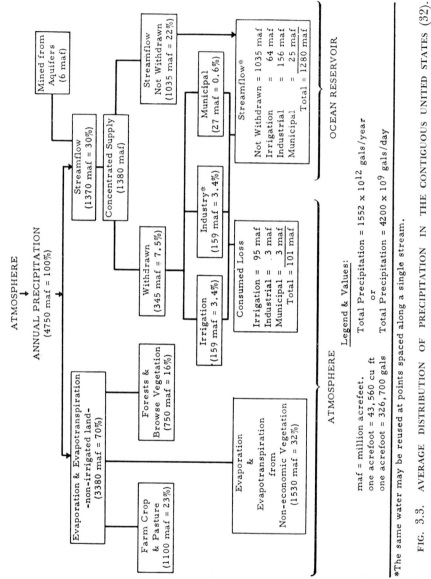

ATMOSPHERE

ANNUAL PRECIPITATION
(4750 maf = 100%)

Mined from Aquifers
(6 maf)

Streamflow
(1370 maf = 30%)

Concentrated Supply
(1380 maf)

Streamflow
Not Withdrawn
(1035 maf = 22%)

Withdrawn
(345 maf = 7.5%)

Municipal
(27 maf = 0.6%)

Industry*
(159 maf = 3.4%)

Irrigation
(159 maf = 3.4%)

Streamflow*

Not Withdrawn = 1035 maf
Irrigation   =   64 maf
Industrial   = 156 maf
Municipal   =   25 maf
          Total = 1280 maf

OCEAN RESERVOIR

Consumed Loss

Irrigation = 95 maf
Industrial =   3 maf
Municipal   =   3 maf
      Total = 101 maf

Evaporation & Evapotranspiration
-non-irrigated land-
(3380 maf = 70%)

Forests &
Browse Vegetation
(750 maf = 16%)

Farm Crop
& Pasture
(1100 maf = 23%)

Evaporation
&
Evapotranspiration
from
Non-economic Vegetation
(1530 maf = 32%)

ATMOSPHERE

Legend & Values:

maf = million acrefeet.
one acrefoot = 43,560 cu ft
one acrefoot = 326,700 gals

Total Precipitation = 1552 x 10^{12} gals/year
          or
Total Precipitation = 4200 x 10^{9} gals/day

*The same water may be reused at points spaced along a single stream.

FIG. 3.3. AVERAGE DISTRIBUTION OF PRECIPITATION IN THE CONTIGUOUS UNITED STATES (32).

may seem extraneous to a discussion of grasslands, it does emphasize the harsh constraints that prevail with respect to benefits for water consumption if crop yield is low. Unfortunately, low yield of plant product from grasslands is altogether too common.

The U.S. census of 1964 revealed that 626 million acres of land in farms were used for pasture, and 292 million acres of land not in farms were used for grazing. Also, 18 million acres were used to produce wild and "other" hay. It is implicit that this latter acreage was largely grasses since legume acreages were listed separately. Grass, therefore, was the key plant product on 926 million acres, one-half the 1900 million acres of land in the conterminous United States.

Our native grasslands are largely in subhumid or semiarid regions so that considerably less than half the 1370 million acre-feet (maf) involved in average annual streamflow and the 3380 maf involved in annual evapotranspiration (Fig. 3.3) is associated with grasslands. For example, the land in the Missouri River basin is mostly natural grassland and that of the Tennessee River basin is natural forest. Table 3.1 shows that even though the mean annual discharge of these two rivers is about the same, the land area in the Missouri River basin is 13 times as great (7, 12, 22). The low average annual precipitation on the Great Plains results in less than 0.5 inch of average annual runoff over much of this great basin (12). In other words, even though most of the precipitation on the Missouri River basin becomes evapotranspiration, the amount is far less than potential evapotranspiration (26) and far less than the 28.5 inches indicated for the Tennessee River basin. Thus these great grasslands are under water deficit much of the time. This limits their productivity and their efficiency in use of water.

TABLE 3.1. *Precipitation and runoff on the basins of the Missouri and the Tennessee rivers*

Precipitation and runoff	Missouri River	Tennessee River
Mean annual discharge (acre-ft)	51,000,000	46,500,000
Drainage area (acres)	339,000,000	26,000,000
Mean annual runoff (in.)	1.8	21.5
Range in mean annual precipitation (in.)	6–44	40–90
Mean annual precipitation on watershed (in.)	19	50

Sources: Chow, 1964 (7); Hart, 1957 (12); Water Resources Policy Comm., 1950 (22).

PRECIPITATION

WATER AND GRASSLANDS are interrelated in terms of various key processes of the hydrologic cycle (Fig. 3.2).

Interception. When precipitation is caught by the vegetative canopy, some of it may not reach the ground surface (17). Conifers will retain appreciable burdens of snow; rain droplets collect in leaves, coalesce, and trickle down the trunks of trees and shrubs and the stems of grasses. Some of this intercepted moisture may actually be absorbed by leaves, or it may evaporate from the surface of the leaf or stems.

Studies by Clark (8) provide quantitative information on the amount of rain that may be intercepted by grasses:

The percentage of interception varied with the intensity of rainfall, density of foliage cover, and environmental conditions. Wind movement and condition of the sky were especially important because of their effect upon evaporation.

Andropogon furcatus [bluestem] intercepted almost 0.5 inch (47 percent) of rain during an hour and larger percentages with applications of lower intensity.

Stipa spartea [speargrass] and *Sporofolus heterolepis* [sand dropseed] withheld 50 percent or more of the water applied in the form of light showers.

Agropyron smithii [western wheatgrass] intercepted almost half and *Elymus canadensis* [Canada wildrye] more than half of a ¼-inch rain during 30 minutes.

Percentage of interception by *Spartina pectinata* [prairie cordgrass] varied from 72 percent with a ⅛-inch rain to 55 with a ½-inch rain during 30-minute periods.

Eragrostis cilianensis [lovegrass] and *Buchloe dactyloides* [buffalograss] prevented from reaching the soil amounts of water ranging from 16 percent during heavy rains to 74 percent during light showers.

Furthermore, the vegetative canopy may be very effective in dissipating the energy of a rainstorm. A 1-inch rain, falling at the rate of 1 inch per hour, imposes nearly 1000 foot-tons of energy on the surface of an acre (31). Vegetative cover is an important factor in dispelling the soil erosivity of a rainstorm. Lull (19) presents a good discussion of interception by vegetation.

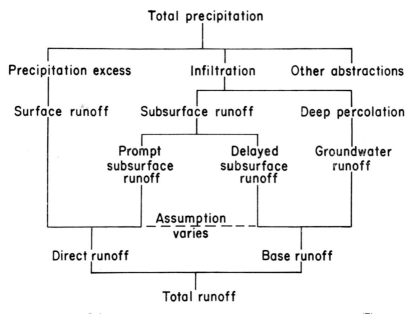

FIG. 3.4.   DESCRIPTIVE TERMS OF RUNOFF PHENOMENA (7).

Surface Disposition. Interest in the fate of precipitation upon the land
is diverse. The farmer, the forester, and the suburbanite desire
the soil moisture reservoir to be recharged by infiltration into the
surface, thereby enabling plants to grow between rainstorms. Those
concerned with use of wells or sustained streamflow are interested in
deep percolation of infiltrated water. Engineers watching dwindling
levels in reservoirs are concerned with water yield and when observing
full reservoirs, they are concerned about floods. Farmers do not want
to see their fields eroded by the hydraulic action of overland flow that
culminates in sediment delivery to stream channels. Water supply
authorities are concerned about reservoirs accumulating sediment.
Downstream water users are becoming more concerned about potential pollutants in land runoff. Infiltration and runoff are indeed
key processes in water disposition on the land.

Infiltration. In simplified form, runoff equals precipitation minus
infiltration, so the rate of infiltration becomes a key determinant
of the nature of water yields downstream. Figure 3.4, Chow (7),
portrays the more detailed relationship between rainfall and runoff.
The effect of infiltration on the precipitation-runoff sequence is

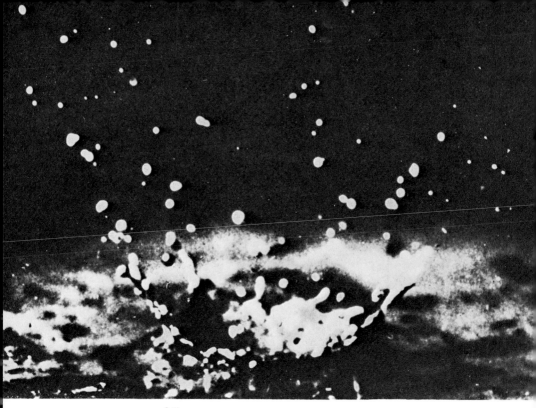

FIG. 3.5. SPLASH FROM A SINGLE RAINDROP.

modified through interception by the vegetative canopy and by undulations on the soil surface that collect puddles and pools which eventually infiltrate or evaporate. Total water yield from the land area includes surface runoff as well as water that infiltrates and becomes subsurface return flow or groundwater accretion. These last two delay and sustain flows, helping to maintain the continuity of flow in streams. The nature of cover on the land is important in determining the course of water depicted in Figure 3.4.

The rate of infiltration into the land surface may range from essentially zero to several inches per hour. The rate will differ greatly among different kinds of soils depending on mineral texture, structure, degree and kind of base saturation, and level of antecedent moisture. Conditions at the surface are critical. Surface water moves into a soil by entering the upper terminals of the least resistive paths through the soil mass. If these upper terminals become clogged with fine particles, rate of infiltration is reduced from the relatively rapid flow through large pores to the much slower process of capillary movement. Thus a single raindrop on a soil surface (Fig. 3.5) can be very disrup-

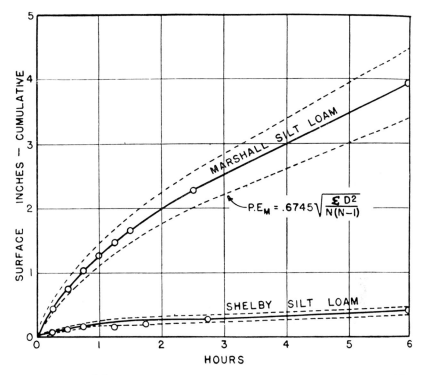

FIG. 3.6.   CUMULATIVE INFILTRATION FOR TWO SOILS OF FIELD
STRUCTURE, BOTH MOIST AND WITH PREVIOUS SOD SURFACE
PARED OFF SMOOTHLY (20).

tive by scattering fine particles which clog pores. A covering of grass
with its associated mulch dissipates the energy of the raindrop and
screens the upper terminals of the noncapillary channels. Rate of infiltration, therefore, is invariably higher under good grass cover than
on barren soil or intertilled crops. Baver (4) presents a good discussion of the factors that affect rate of infiltration.

Figure 3.6 is a plotting of Musgrave's observations (20) on the
marked difference in infiltration of a Shelby loam at Bethany, Mo.,
and a Marshall silt loam at Clarinda, Iowa. Infiltration into the
Marshall soil was nearly eight times that into the Shelby soil. Baver
(4) found that noncapillary porosity is about five times higher in the
Marshall than in the Shelby (Fig. 3.7). For infiltration to take place
voids must exist within the soil into which water can move, and
the surface openings to the porosity of the soil mass must not become
clogged.

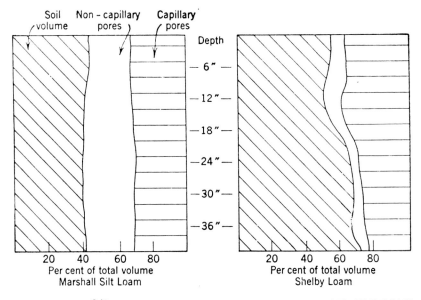

FIG. 3.7.   PORE-SPACE RELATIONSHIPS IN MARSHALL SILT LOAM
AND SHELBY LOAM (4).

FIG. 3.7.   PORE-SPACE RELATIONSHIPS IN MARSHALL SILT LOAM
AND SHELBY LOAM (4).

The USDA Hydrograph Laboratory has developed a model of
infiltration that takes into account the noncapillary and capillary
porosities of soils predominant on a watershed. This approach was
essential in developing an analytical approach to the hydrology of
agricultural watersheds.

The equation derived from logarithmic plottings of data from
plots, small watersheds, and infiltrometer runs is as follows (15):

$$f = AS_a^{1.4} + f_c$$

where f = rate of infiltration, in inches per hour
     f_c = rate of infiltration after prolonged wetting, in inches per
         hour
     S_a = available soil porosity undepleted by moisture at any
         given time, in inches of water depth
     A = a coefficient of the pore-space continuity estimated as the
         product of vegetative density at crop maturity and stage of
         growth in percent

The course of infiltration capacity generally follows the curve in
Figure 3.8. This rate usually starts at a maximum, decreases expo-

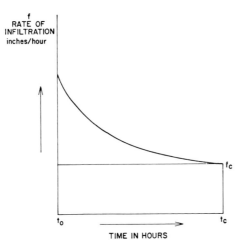

FIG. 3.8.   SCHEMATIC RELATION OF INFILTRATION AND TIME.

nentially, and approaches the stable value, f_c, asymptotically. Further symbols may now be defined:

t_o = start of infiltration
t_c = time at which f reaches the stable rate shown as f_c
S_o = available soil porosity undepleted by moisture at t_o, in inches of water depth
S_c = available soil porosity, in inches at t_c
S_f = $S_o - S_c$, the depth of water in inches that is infiltrated between t_o and t_c
$(S_f/S_o) \times 100$ = percentage of available porosity filled at t_c

With the data in Figure 3.9 Holtan (15) shows that density of vegetative cover is a good predictor of the proportion of available porosity that is filled when rate of infiltration has decreased to a stable rate. This approach enables quantifying the beneficial effects of grass and other types of cover in the development of a mathematical model to predict the behavior of watersheds and the protective treatment that may be needed.

Holtan and Kirkpatrick (16) present a summary chart (Fig. 3.10) of the relative effect of surface cover on rate of infiltration. The chart emphasizes that grass cover invariably affords more infiltration than any other agricultural treatment affecting the soil surface. Grass roots aid in maintaining porosity in a soil, and the surface mulch protects the surface from the action of raindrops in dispersing

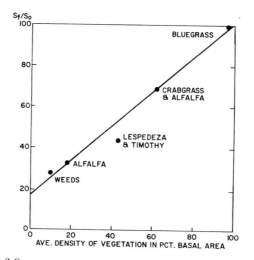

FIG. 3.9. RELATION OF DENSITY OF VEGETATION TO PERCENT-
AGE OF AVAILABLE POROSITY FILLED WHEN INFILTRATION
REACHES A STABLE RATE (15).

soil and clogging pores. Rainfall that enters into the soil does not become surface runoff. It is of immediate interest then to note how the
superior infiltration observed under grass cover affects surface runoff.

Runoff. Between about 1930 and 1961, the USDA in cooperation with
state agricultural experiment stations established 1231 runoff

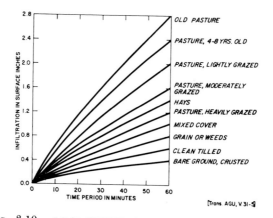

FIG. 3.10. MASS INFILTRATION ON CECIL, MADISON, AND DUR-
HAM SOILS BASED ON INFILTROMETER DATA (16).

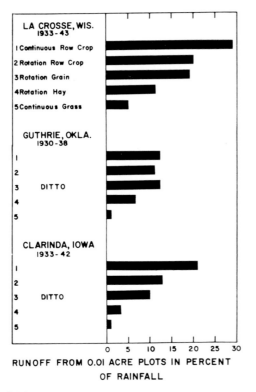

FIG. 3.11.   EFFECT OF FARM PRACTICES ON RUNOFF VOLUMES
(11).

plots ranging from 0.01 to 1.0 acre in size on 45 different soils and
under a variety of climatic conditions.

Figure 3.11 (11) shows the general trend of results obtained at
three latitudes in the central United States by comparing runoff
amounts from continuous row crops, crops in three-year rotations, and
continuous grass. At the Upper Mississippi Valley Conservation Experiment Station, La Crosse, Wis., average annual runoff on Fayette
silt loam with a slope of 16 percent was 27.7 percent of rainfall for
continuous corn; 20.6 percent for corn in rotation of corn–barley–red
clover; 18.9 percent for barley in rotation; 11.5 percent for clover in
rotation; and 5.5 percent for continuous bluegrass. Results were
similar for the Red Plains Conservation Experiment Station, Guthrie,
Okla., and the Missouri Valley Loess Conservation Experiment Station, Clarinda, Iowa. It is evident that runoff was inversely related to

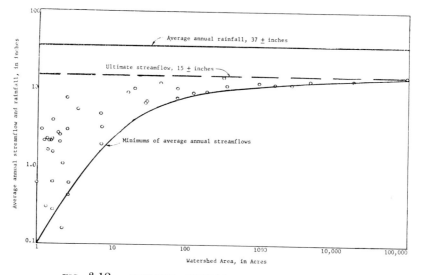

FIG. 3.12. AVERAGE ANNUAL STREAMFLOW FROM AGRICUL-
TURAL WATERSHEDS IN THE VICINITY OF COSHOCTON, OHIO.

rate of infiltration and that density of vegetation and the frequency of cultivation were determinative factors.

The decrease of runoff under grass as compared to tilled crops does not imply that water yield to streams will also decrease. Plot and watershed studies by the Agricultural Research Service near Coshocton, Ohio, revealed drastic differences in surface runoff from small upland areas as indicated by the scatter of points in Figure 3.12. The 20-year average annual outflow varied from 15 inches per year to less than 1 inch per year on small areas planted to hays and grasses, but the average outflows for the larger areas were all near 15 inches per year. Since none of the small areas yielded more than 15 inches per year, we may conclude that downstream yields in this case are sustained by subsurface return flow from upland infiltration. Hydrograph analysis confirms this premise, since groundwater return flow is recognizable as a prolonged flow recession in the hydrograph. Assuming that free water will flow laterally as at Coshocton, or downward to the water table as observed in lysimeters and in groundwater studies, consumptive use by plants seems to be our best estimate of water yield reductions due to land use.

The potential of grass and associated conservation practices to affect peak rates of runoff from watersheds at Hastings, Nebr., is shown in Figure 3.13 (3). The 481-acre watershed was farmed in

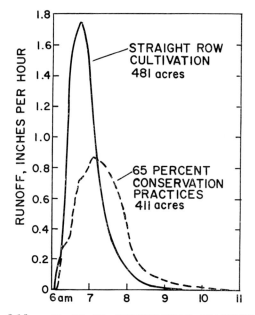

straight rows of intertilled crops. The 411-acre watershed was one-third grasslands and the remaining two-thirds was cropped and contained terraces and grassed waterways. For the intense storm of July 10, 1951, peak flow on the unprotected watershed was double that from the watershed under conservation treatment. During the two-day period, July 10–12, the treated watershed retained 0.6 inches more of the approximately 5 inches of rain than did the unprotected watershed. Unlike the experience at Coshocton, the geologic stratigraphy at Hastings does not cause return flow, and thus free water from infiltration flows downward to groundwater.

Character of Watershed. The U.S. Geological Survey (35) has presented a striking comparison between two grassed watersheds to illustrate how the character of the watershed affects subsurface storage, subsurface flow, and continuity of flow for a stream. Table 3.2 presents pertinent information about the basins of the Middle Loup River that flows through the sandhills of Nebraska, and the Bad River that flows through the Pierre shales of South Dakota.

The Valentine sands of Nebraska have very high natural infil-

TABLE 3.2.   *Some comparisons of the Middle Loup and Bad River basins*

Item	Middle Loup River (west central Nebraska)	Bad River (south central South Dakota)
Predominant soils	Valentine sand	Pierre clay loam
Vegetative cover	short and tall prairie grasses	short grasses and forbs
Geologic substrata	sand	impervious shale
Drainage area (mil acres)	3.04	1.98
Mean annual precipitation (in.)	20	15
Mean annual streamflow (000 acre-ft)	584	128
Mean annual runoff (in.)	2.31	0.78
Percent runoff	11.6	5.2
Minimum mean daily flow (ft³/sec)	92	0
Maximum mean daily flow (ft³/sec)	1500	2750
Flow available 50% of time (ft³/sec)	900	7

Source: U.S. Senate, 1960 (35).

tration capacity with little runoff except during very severe storms. Appreciable amounts of infiltrated water percolate downward to the water table. Subsurface lateral flow of this water to the stream channel results in remarkable uniformity of flow in the Middle Loup River (Fig. 3.14) (35). The Bad River flows through a basin formed largely from Pierre shales. Although this basin is mainly rangeland, as is that of the Middle Loup, the vegetative cover is not as good. The Pierre soils are essentially impervious to water, except when they dry to the point where tremendous cracks occur. During rainfall these cracks and undulations in the soil surface hold appreciable amounts of water, but rate of infiltration into the soil is low. A heavy rain will cause high surface runoff but little subsurface movement of water to the stream channel. Thus the Bad River is often dry, and rate of flow varies widely (Fig. 3.14). A storm that induces runoff may produce very high flows with no sustenance from delayed return flow. This comparison between the two basins emphasizes the importance of good infiltration and storage capacity of watershed soils plus subsurface transfer to a stream channel to maintain stability of streamflow.

Hibbert (14) obtained remarkable evidence of the increase in water yield that results from converting subhumid watersheds in Arizona from chaparral brush to grass. Figure 3.15 shows that eliminating brush with aerial applications of 2,4,5-T and fostering growth of grass increased water yield 1–4 inches per year. Deep-rooted shrubs depleted the soil moisture reservoir more than did the grasses, result-

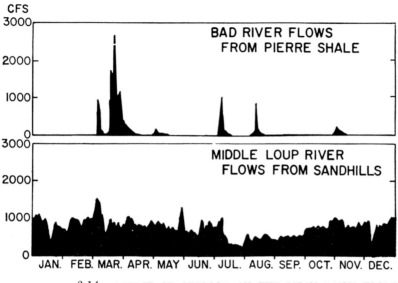

FIG. 3.14. EFFECT OF GEOLOGY ON THE MEAN DAILY FLOWS OF THE BAD AND MIDDLE LOUP RIVERS (35).

FIG. 3.15. CONTROL AND TREATMENT REGRESSIONS SHOWING TREATMENT EFFECT (DIFFERENCE BETWEEN LINES) ON WATERSHED B. LINES FITTED BY LINEAR REGRESSION AFTER LOGARITHMIC TRANSFORMATION OF THE DATA (14).

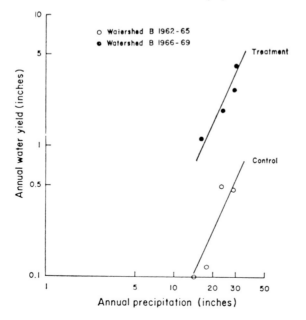

ing in greater retention of the winter rains that provide most of the streamflow.

These brief comments emphasize the excellent attributes of good grass cover in beneficially modifying land surfaces for desirable water behavior on the land. Additional benefits include the protection of the soil surface and the reduced surface runoff under grass that are also associated with great reductions in soil loss and sediment delivery.

Saxton (private communication) has provided data from paired watersheds at Treynor, Iowa, from cooperative hydrological research between the Agricultural Research Service and the Iowa and Missouri agricultural experiment stations. One 75-acre watershed is maintained in continuous corn without terraces or other conservation treatments. The other, a grassed watershed of 103 acres, also receives no conservation treatment other than maintenance of fairly good grass cover. Figure 3.16 shows the comparison in hydrologic responses for one storm on June 9, 1967. The amount of precipitation was nearly the same on each watershed. Recording rain gages revealed minor differences in the precipitation patterns even though the watersheds were located within the same square mile. Sediment from the grassed area was only about 0.01 percent of that from the corn. The effectiveness of good grass cover in preventing soil erosion and sediment delivery is outstanding in all studies on land runoff.

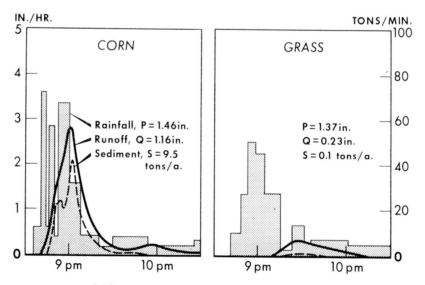

FIG. 3.16.   IMPACT OF LAND USE UPON HYDROLOGIC RESPONSE OF TWO WATERSHEDS, TREYNOR, IOWA, JUNE 9, 1967.

TABLE 3.3. *Effect of cropping on soil erosion*

Location	Treatment	Annual soil loss
		(tons/acre)
La Crosse, Wis.	continuous row crop	111.7
	rotation row crop	52.9
	rotation, grain	29.8
	rotation, hay	0.68
	continuous grass	0.1
Guthrie, Okla.	continuous row crop	33.0
	rotation row crop	3.2
	rotation, grain	6.9
	rotation, hay	2.5
	continuous grass	0.02
Clarinda, Iowa	continuous row crop	36.9
	rotation row crop	18.4
	rotation, grain	10.1
	rotation, hay	5.4
	continuous grass	0.03

Sources: Browning et al., 1948 (5); Daniel et al., 1943 (9); Hays et al., 1949 (13).

Protective Action of Grass. Table 3.3 presents data (5, 9, 13) on soil loss for the same field-plot studies used to develop Figure 3.11.

The protective action of grass is so evident that it is not even necessary to call in the statisticians. This protective action illustrates why grassed waterways are used in conservation planning on farms to provide the most economical means of conveying surface runoff from fields without incurring erosion of the channel. The grassed water-way may at times be a minor nuisance in tillage operations, but a gully through an unprotected draw would be far worse.

Table 3.4 shows summary information on the water flow that may be permissible over various types of grass cover. The data are a small part of the information developed at the Stillwater Outdoor Hydraulic Laboratory, Stillwater, Okla., operated cooperatively by the USDA and the Oklahoma Agricultural Experiment Station (33).

The protective action of grass in maintaining good infiltration, in curbing runoff, and in nearly eliminating sediment delivery has important implications in the transport of potential pollutants from the land to surface waters. In terms of total mass, sediment itself is the dominant water pollutant (30). Something in excess of 4 billion tons of sediment move from the land to water channels during the average year. If chlorinated hydrocarbon pesticides move from fields to water, sediment is the carrier (34). These chemicals are virtually insoluble in water but are readily adsorbed on the surfaces of clay particles. Phosphate rides by surface adsorption on sediment moving from fields to streams. Soluble phosphate is invariably at low concen-

TABLE 3.4. *Permissible velocities for channels lined with grasses*

Cover	Slope range	Permissible velocity (ft/sec)	
		Erosion-resistant soils	Easily eroded soils
	(%)		
Bermudagrass	up to 5	8	6
	5–10	7	5
	over 10	6	4
Buffalograss Kentucky bluegrass Smooth bromegrass Blue grama	up to 5 5–10 over 10	7 6 5	5 4 3
Sericea lespedeza Weeping lovegrass Yellow bluestem Kudzu Alfalfa Crabgrass	up to 5	3.5	2.5
Common lespedeza Sudangrass	up to 5	3.5	2.5

Source: USDA, 1954 (33).

tration, 0.01–0.3 ppm, in runoff water. Manure particles ride from fields, barnyards, or feedlots to streams by the same hydraulic forces that move sediment. It is implicit, therefore, that grassed areas are tremendously important in watershed protection and pollution abatement by projects under Public Law 83-566 (the Watershed Protection and Flood Prevention Act) and similar programs.

Studies (28) at the Campbell Soup Company plant at Paris, Tex., provide clear evidence of the exceedingly remarkable effectiveness of grass in rectifying polluted water at low cost. The Paris canning plant discharges up to 3.6 million gallons of water a day. This water is contaminated from use in the preparation of food products and contains from 550 to 900 ppm biochemical oxygen demand (BOD). Thus the effluent from the Paris plant is about three times as polluted as raw city sewage. Paris, Tex., is located in the Houston-Austin soils association. The soil is a clay loam that has been ravaged by severe sheet and gully erosion while under row crops. The soil is so erodible that it always should have had grass cover. This impervious soil has an infiltration rate of 0.10 inch per day or less when swelling has eliminated the major cracks that may occur on drying. Obviously, runoff from rainfall or irrigation is exceedingly high. Some 400 acres of land were smoothed with a land plane and terraced at 200- to 300-foot intervals. The smoothed areas were planted to grass after a sprinkler irrigation system with underground mains was installed along the upper side of the smoothed area between the terraces. On establishment of the grass cover, the canning plant effluent was applied at the rate of about 0.1 inch per hour.

Since the soil had exceedingly low infiltration capacity, 80–90 percent of the applied polluted effluent became runoff at the lower terrace of each smoothed area. Even though most of the polluted water did not enter the soil but was intercepted by the grass and trickled slowly downslope on the soil surface, 99 percent of the BOD was removed by the time the water reached the lower terrace. Up to 90 percent of the phosphorus and nitrogen in the polluted effluent was removed by the time the tricklings reached the lower terrace.

This field installation is a truly remarkable demonstration of the tremendous capability of grass to rectify seriously polluted water. We are informally advised that the cost for this rectification procedure is exceptionally modest.

Studies on the hay produced by the reed canarygrass grown with this polluted water showed that it was relatively high in total digestible nutrients. When this hay was fed to cattle having the optional choice of three other kinds of hay of good quality, the cattle showed a preference for the reed canarygrass hay grown with polluted water.

It is our view that the potential for the use of grass in environmental protection is just beginning to be appreciated.

EVAPOTRANSPIRATION

As SOLAR RADIATION reaches the earth's surface, some is absorbed by the surface and some is reflected back to the sky; that absorbed is termed net radiation. The heat budget in the net radiation may be used to evaporate water, heat plants or other entities on the soil surface, heat the soil, and heat the air. One percent or less will be used in photosynthesis. Tanner (25) illustrates this by the diagram in Figure 3.17, i.e., net radiation (R_n) = soil heat (S) + heat used in evapotranspiration (E) + sensible heat (A) (air) + energy used in photosynthesis (P).

Penman (21) emphasizes that when a crop is adequately supplied with water at times of the year when net radiation and evapotranspiration are large enough to be hydrologically important, the ratio of $(S + A + P)/E$ is small (0.1 to 0.2 for humid climates). Thus the

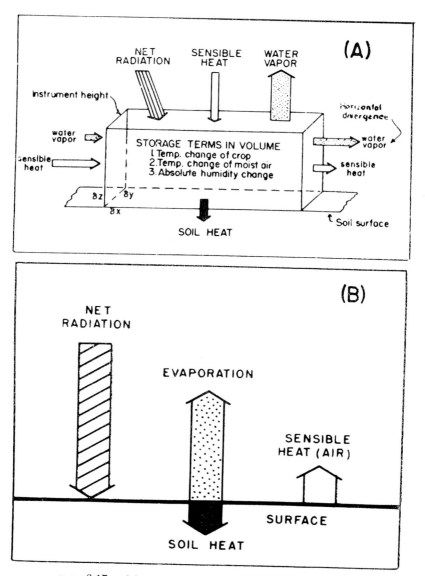

FIG. 3.17. (A) COMPLETE ENERGY BALANCE OF A CROP VOL-
UME. (B) VERTICAL ENERGY BALANCE OF A CROPPED SURFACE
(25).

value of E is determined almost entirely by the magnitudes of R_n. Hence evapotranspiration is largely independent of the kind of crop since E is determined predominantly by meteorological conditions. It is implicit from Penman's analysis that little can be done through selection of crops or crop management to improve efficiency of water use by resorting to some particular technique to artificially curb evapotranspirational losses. An obvious corollary to this implication is that water use efficiency will be increased as the small proportion of net radiation used in photosynthesis is increased to bring about an increase in net assimilation in the crop per unit area of land surface. The latter is the sort of approach being emphasized by Duncan (10) and Loomis et al. (18). Burton (6) has emphasized the objectives of improving water use efficiency of bermudagrass by selection of strains relatively high in net assimilation per unit loss of water in transpiration.

Grasses are indigenous to the prairies and plains where potential evapotranspiration appreciably exceeds actual evapotranspiration over much of the growing season and a net water deficit exists for the year (27). Thus it is exceedingly important for grass species on most of our grasslands to make some growth or even merely to survive under increasing intensities of soil moisture stress. As soil water becomes limiting to meet evaporative potential of the net radiation, the temperature of grass leaves will rise and tissues will be desiccated. Tolerance of these stresses by grasses providing land surface protection and animal feed is essential.

Figures 3.18 and 3.19 (29) show the average trend of potential evapotranspiration and rainfall distribution over the year for New England; N. Dak.; Tahoka, Tex.; and northern and southern locations in the Great Plains.

In years of average precipitation, there is wide disparity during the growing season between potential evapotranspiration and the amount of rainfall available for actual evapotranspiration. The relatively low storage of soil moisture that may be expected during late fall and winter is quite inadequate to make up the deficiency. Even in the 25 percent of years having the most precipitation, a serious moisture deficit exists on the Great Plains during the growing season. The capability of rangeland grass to tolerate and survive the serious climatic stresses that prevail becomes fundamental in the protection of soil and water resources in subhumid and semiarid regions.

Smika et al. (24) have studied consumptive use of water by native range grasses, predominantly blue grama, western wheatgrass, and

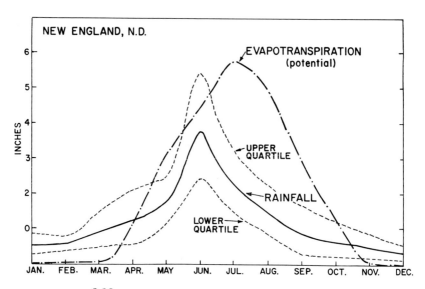

FIG. 3.18. AVERAGE ANNUAL RAINFALL AND POTENTIAL EVAP-
OTRANSPIRATION, NEW ENGLAND, N. DAK. (29).

FIG. 3.19. AVERAGE ANNUAL RAINFALL AND POTENTIAL EVAPO-
TRANSPIRATION, TAHOKA, TEX. (29).

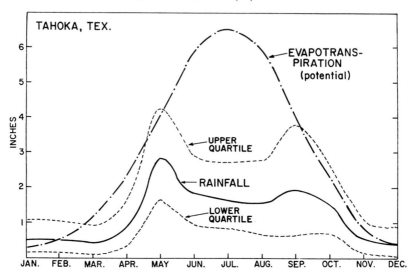

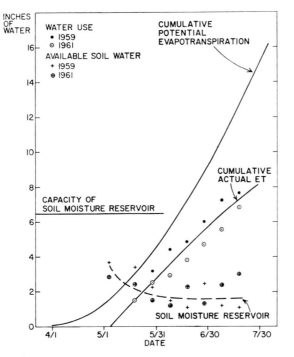

FIG. 3.20. EVAPOTRANSPIRATION AND SOIL MOISTURE, MAN-
DAN, N. DAK. (24).

needle-and-thread, at Mandan, N. Dak. They provided supplemental
information to enable presentation of Figure 3.20. The soils at that
location have a potential storage capacity of 6.5 inches of available
soil moisture to the 6-foot depth. The data for evapotranspiration
and soil moisture storage are averages for the years 1959 and 1961
and are quite similar. The curve of potential evaporation at Mandan
was calculated from weather records.

At the beginning of each of the three seasons, the soil moisture
reservoir was little more than half filled. As the season progressed,
the divergence between potential and actual evapotranspiration became progressively wider. The soil moisture reservoir was seldom
more than two-thirds filled during the study period. These data emphasize the need for grass to be highly tolerant to heat and evapora-
tional stresses on the Great Plains.

Savage and Jacobson (23) have documented the heavy toll, even
on heat- and drought-tolerant grasses, when available soil moisture
is seriously inadequate in terms of potential evapotranspiration. A

nearly continuous and disastrous drought prevailed at Hays, Kans., from October 21, 1932, to August 30, 1934. It was the hottest and driest biennium since 1894–1895. The effect on native short grasses of the Plains was described as follows:

The average percentage of short grasses killed by the heat and drought of 1933–34 was 74.8 on closely grazed and severely trampled areas, 64.6 on moderately grazed areas, and 44.4 on unwatered lawns. Repeated applications of water to lawns in 1934 were decidedly beneficial in overcoming the drought of 1933 and counteracting similar conditions in 1934. Only 14.5 percent of the short grasses failed to survive on the lightly watered lawns and only 5.1 percent on the heavily watered lawns.

SUMMARY

THE CAPACITY of good grass cover to disperse the entry of rainfall, to maintain maximal infiltration capacity of soil, to minimize runoff, to nearly eliminate sediment delivery, to rectify polluted water, and to tolerate and survive serious soil moisture stress so as to provide continual protection of soil and water resources makes grasses of fundamental importance in many undertakings for improved environmental protection.

REFERENCES

    1. Ackerman, E. A., and G. O. G. Lof. 1959. Technology in American Water Development. Johns Hopkins Press, Baltimore.
    2. Ackermann, W. L., E. A. Colman, and H. O. Ogrosky. 1955. From ocean to sky to land to ocean. *In* Water, pp. 41–51. USDA Yearbook Agr.
    3. Allis, John A. 1952. The story of two watersheds. *J. Soil and Water Conserv.* 7 (5):243–45.
    4. Baver, L. D. 1959. Soil Physics. Wiley, New York.
    5. Browning, G. M., R. A. Norton, A. G. McCall, and F. G. Bell. 1948. Investigation in erosion control and the reclamation of eroded land at the Missouri Valley Loess Conservation Experiment Station, Clarinda, Iowa, 1931–42. USDA Tech. Bull. 973.
    6. Burton, G. W. 1959. Soil-plant relationships: Crop management for improved water use efficiency. *Adv. Agron.* 11:104–10.
    7. Chow, V. T. 1964. Runoff. *In* Handbook of Applied Hydrology, pp. 1–54. McGraw-Hill, New York.
    8. Clark, O. R. 1940. Interception of Rainfall by Prairie Weeds, Grasses, and Certain Crops Plants. Ecol. Monogr. 10:243–77.
    9. Daniel, H. A., H. M. Elwell, and M. B. Cox. 1943. Investigations in erosion control and reclamation of eroded land at the Red Plains Conservation Experiment Station, Guthrie, Oklahoma, 1930–40. USDA Tech. Bull. 837.

10. Duncan, W. G. 1967. Model building in photosynthesis. *In* Harvesting the Sun, pp. 309–14. Academic Press, New York.

11. Glymph, L. M., and H. N. Holtan. 1969. Land treatment in agricultural watershed hydrology research. *In* Effects of Watershed Changes on Streamflow, pp. 44–68. Univ. Texas Press, Austin.

12. Hart, H. L. 1957. The Dark Missouri, p. 260. Univ. Wis. Press, Madison.

13. Hays, O. E., A. G. McCall, and F. G. Bell. 1949. Investigations in erosion control and the reclamation of eroded land at the Upper Mississippi Valley Conservation Experiment Station near La Crosse, Wisconsin, 1933–43. USDA Tech. Bull. 973.

14. Hibbert, A. R. 1971. Increases in streamflow after converting chaparral to grass. *Water Resources Res.* 7:71–80.

15. Holtan, H. N. 1961. A concept for infiltration estimates in watershed engineering. USDA Agr. Res. Ser., pp. 41–51.

16. Holtan, H. N., and M. H. Kirkpatrick, Jr. 1950. Rainfall, infiltration, and hydraulics of flow in runoff computation. *Trans. Am. Geophys. Union* 31:771–79.

17. Horton, R. E. 1931. The field, scope, and status of the science of hydrology. *Trans. Am. Geophys. Union* 12:189–202.

18. Loomis, R. S., W. A. Williams, and W. G. Duncan. 1967. Community architecture and the productivity of terrestrial plant communities. *In* Harvesting the Sun, pp. 291–308. Academic Press, New York.

19. Lull, H. W. 1964. Ecological and silvicultural aspects. *In* Handbook of Applied Hydrology, pp. 1–30. McGraw-Hill, New York.

20. Musgrave, G. W. 1935. The infiltration capacity of soils in relation to the control of surface runoff and erosion. *J. Am. Soc. Agron.* 27:336–45.

21. Penman, H. L. 1963. Vegetation and hydrology. Commonwealth Bureau of Soils. Harpenden Tech. Comm. 13:124.

22. President's Water Resources Policy Commission, The. 1950. Ten Rivers in America's Future, Vol. 2. USGPO.

23. Savage, D. A., and L. A. Jacobson. 1935. The killing effect of heat and drought on buffalo grass and blue grama grass at Hays, Kansas. *J. Am. Soc. Agron.* 27:566–82.

24. Smika, D. E., H. J. Hass, and J. F. Powers. 1965. Effects of moisture and nitrogen fertilizer on growth and water use by native grass. *Agron. J.* 57: 483–86.

25. Tanner, C. B. 1960. Energy balance approach to evapotranspiration from crops. *Soil Sci. Soc. Am. Proc.* 24:1–9.

26. Thornthwaite, C. W. 1954. A reexamination of the concept and measurement of potential evapotranspiration. *Climatology* 7:200–209.

27. Thornthwaite, C. W., J. R. Mather, and D. B. Carter. 1958. 3 water balance maps of eastern North America. Johns Hopkins Press, Baltimore. (Publications for Resources for the Future, Inc.)

28. Thornthwaite, C. W., et al. 1969. An evaluation of cannery waste disposal by overland flow spray irrigation, Campbell Soup Company, Paris [Texas] plant. *Climatology* 22 (2):1–74.

29. Wadleigh, C. H. 1961. Problems of arid areas, pp. 157–180. 1960 Western Resources Conference, Univ. Colo. Press.

30. ———. 1968. Wastes in relation to agriculture and forestry. USDA Misc. Publ. 1065.

31. Wischmeier, W. H., and D. D. Smith. 1958. Rainfall energy and its relationship to soil loss. *Trans. Am. Geophys. Union* 39:285–91.

32. Wolman, A. 1962. Water resources, a report to the committee on natural resources of the National Academy of Sciences. National Research Council Publ. 1000-B.

33. USDA. 1954. Handbook of channel design for soil and water conservation. SCS Publ. TP-61.

34. USDA. 1969. Monitoring agricultural pesticide residues 1966–67. A final report on soils, crops, water, sediment, and wildlife in six study areas, p. 97. ARS 81-32.

35. United States Senate. 1960. Water resource activities in the United States. Select Committee on National Water Resources, 86th Congr., 1st sess. Committee Print 3.

4

IMPORTANCE OF FORAGES
TO LIVESTOCK PRODUCTION

H .   J .   H O D G S O N

SALES OF LIVESTOCK and livestock products in the United States were valued at almost $22 billion in 1965 (13). This represents about 56 percent of total sales of agricultural commodities. Sales of products produced by cattle and sheep were valued at $14.5 billion, about two-thirds of all livestock products. Cattle and sheep are capable of converting plant materials not usable to any considerable degree by most nonruminant animals, including man, to forms that can be utilized by and are highly palatable to man. Historically, the ruminant, although entirely capable of utilizing large amounts of concentrates, has depended upon forages and roughages as the principal source of feed.

USE OF FEED GRAINS

IN THE past quarter-century, cereal grains have become important sources of feed for ruminants in the United States. During this period application of highly sophisticated production technology to genetically superior cereal crop varieties resulted in production quantities greatly in excess of the needs of man and nonruminant animals. Mounting surplus production resulted in depressed relative prices for cereal crops, and they became increasingly attractive and valuable as sources of feed for ruminants in the production of meat and dairy products. Feed grains are rich energy sources and, by their physical

Reprinted with permission from ASA Special Publication 13, Forage economics in quality, pp. 11–25, 1968. Published by the American Society of Agronomy, Crop Science Society of America, and Soil Science Society of America.

nature, are well suited to the mechanized and automated feeding that has accompanied, and perhaps made possible, increased size of production units in segments of the livestock industry.

Large quantities of feed grains are fed to livestock. About 8 percent of the corn crop, most of the oats, and a large part of the grain sorghums and barley are so used. This averaged about 118 million tons in the 1961–65 period (15). Wheat is also used as a feed grain, the amount depending somewhat upon the relationship between production and human food requirements and, therefore, relative prices.

Many agriculturalists (including agricultural administrators, farm magazine editors, livestock producers, agronomists, and others) believe that feed grains provide the major portion of feed units consumed by cattle and sheep. In this connection it is especially enlightening to examine data compiled in various reports by the Economic Research Service, USDA, to provide an estimate of actual value of forages in livestock production.

USE OF FORAGES

ALLEN AND DEVERS (1) list tonnages of feed and feed units derived from various sources consumed in 1965 by all livestock, by ruminant animals (cattle, sheep, and goats) and by horses and mules. These data are given in Table 4.1; consumption of feed units expressed in percentages is given in Table 4.2.

Data for hogs and poultry are not included in these and subsequent tables except as they are included in data for all livestock. Forages are minor constituents of rations for these animals. A feed unit is defined as the feed value equivalent of 1 pound of corn. Concentrates include feed grains, high-protein feeds, and other by-product feeds. Forages include hay, other harvested forages (silage, stover, straw), and pasture, including range.

These data show that forages accounted for slightly more than half the feed units consumed by all livestock. Dairy cattle obtained two-thirds and beef cattle three-fourths of their feed units from forages. Sheep, goats, horses, and mules obtained much larger proportions of their feed units from forages, although total feed consumption by these animals is relatively minor. The importance of forages in the production of beef, dairy products, and mutton and wool is evident. However, the extent of dependence of this segment of the livestock industry on forages is not generally realized.

TABLE 4.1. *Consumption of feed and feed units by kind of livestock, 1965*

Livestock class	Concentrates		Hay		Other harvested forage		Pasture	Total forage	Total feed
	Feed	Feed units	Feed	Feed units	Feed	Feed units	Feed units	Feed units	Feed units
					(mil tons)				
All livestock	163.7	174.0	119.4	48.9	123.4	22.3	131.3	202.5	376.6
Dairy cattle	28.6	29.7	58.8	23.8	73.5	11.9	21.4	57.2	86.9
Beef cattle	33.3	35.2	54.8	22.7	45.0	8.9	94.4	126.1	161.4
Sheep and goats	.9	1.2	.9	.4	2.2	.4	9.2	10.3	11.2
Horses and mules	1.6	1.5	3.4	1.4	2.7	.7	3.2	5.4	6.9

Source: Allen and Devers, 1967 (1).

TABLE 4.2. *Percentage of feed units consumed by different classes of livestock, derived from different sources, 1965*

Livestock class	Concentrates	Hay	Other harvested forage	Pasture	All forage	Total
All livestock	47	13	5	35	53	100
Dairy cattle	34	27	14	25	66	100
Beef cattle	22	14	6	58	78	100
Sheep and goats	11	3	4	82	89	100
Horses and mules	22	21	10	47	78	100

Note: Computed from data in Table 4.1.

MONETARY VALUE OF FORAGES

IT IS DIFFICULT to attempt to place a monetary value on forage crops because they are marketed through livestock, often on the farm where they are produced, and therefore appear indirectly in cash receipts as a return from animal products marketed. Also, census figures on forage production are very inadequate when compared with feed grains, for example.

However, this problem can be approached indirectly, and dollar values can be estimated by several methods. Table 4.3 provides an estimate of dollar values for forages resulting from the different methods of calculation (13). Values range between about $7 and $8 billion. Method A simply totals all forage feed units for all livestock (Table 4.1) and equates this to the same number of pounds of corn valued at $40 per ton. This method, which produces the highest value, is subject to the criticism that a considerable portion of the feed units involved (particularly those of pasture and range), although having the equivalent feed value of a pound of corn, are actually of less cash value because of physical location and because they actually could not be marketed. Nevertheless, the fact remains that these feed units are consumed and marketed as animal product and therefore have value. Similarly, the 85 percent of our corn crop fed to livestock has few other markets and would have much reduced value except for the feed market.

Method B estimates dollar values by assigning values per ton to hay, silage, and pasture (or hay equivalent). Values of these items vary from season to season and in different parts of the country. It appears, however, that the assigned values are realistic averages and the dollar value estimate for all forage consumed is not unduly biased.

Method C employs a considerably different method of calculation

TABLE 4.3. *Dollar values of forages estimated by three methods*

METHOD A: Tons of forage feed units consumed by livestock multiplied by corn price per ton
202.5 million @ $40 per ton = $8.1 billion

METHOD B: Based on estimated value per ton of forages consumed

Hay	119.4 million tons @ $25 =	$2.985 billion
Other harvested forage	123.4 million tons @ 10 =	1.234 billion
Pasture*	328.0 million tons @ 10 =	3.280 billion
	Total forage value =	$7.499 billion

METHOD C:

	Receipts as feed costs	Feed units as forage	Cash receipts as forage value	Cash receipts†	Forage dollar value
		(%)		(mil $)	
Beef cattle‡	70	78	54.6	$6507	$3553
Dairy cattle	50	66	33.0	5070	1673
Sheep	70	89	62.3	341	212
Hogs, poultry, horses (estimated forage value)					300
			Total forage value		$5738

* Converted to hay equivalent using factor of 0.4 in relation to corn feed unit of 1.0.
† Cash receipts reported in USDA, 1966 (13). Receipts for beef cattle reduced by $2400 million, the amount paid for feeder cattle in 1965.
‡ Includes dairy animals slaughtered for beef.

and involves several assumptions subject to error. This method is based on cash receipts for livestock and livestock products and involves making an estimate of the percent of cash receipts that represents feed costs for each class of livestock. This percentage multiplied by the percent of feed units furnished by forages identifies a portion of cash receipts that represents the forage contribution to each class of livestock. This method provides the lowest estimate of forage dollar value. It does not take into account residual forage values in about 60 million head of beef animals not marketed. Only about one-fourth the beef cattle on farms in any one year is marketed.

A further criticism of this method is that it uses feed costs in livestock production rather than feed value, which is quite different. For example, in the case of beef cattle 30 percent of cash receipts are apportioned to the animal, labor, land costs, housing, return on investment, management, etc. While these are real costs to the livestock producer, they are paid for by the feed-animal combination and the value of the feed is accordingly somewhat higher than its cost. It is believed, therefore, that this method considerably underestimates the true value of forages in livestock production. Of the three methods

TABLE 4.4. *Value in 1965 and 1966 of major field crops*

Crop	1965	1966
	(bil $)	
Corn*	$4.800	$4.920
Cottonseed and lint	2.354	1.579
Soybeans	1.940	2.524
Wheat	1.603	2.025
Tobacco	1.186	1.210
Rice	0.374	0.405
Potatoes	0.781	0.587
Sugar beets	0.237	0.245
Peanuts	0.279	0.270

Note: Values except for corn are cash receipts taken from USDA, 1966 (13).
* Value for corn calculated from USDA, 1967 (13).

of estimation, method A is probably the most accurate. Comparison of calculated forage crops values with cash receipts for certain other crops for which reasonably accurate data are available is of interest. Table 4.4 lists cash receipts for other major crops for 1965 and 1966 (13). As can be seen readily, the average calculated value of forage derived in Table 4.3 almost equals the combined values of cotton, soybeans, wheat, tobacco, and rice.

SOURCES OF FEED UNITS

THE COMPOSITION of "other harvested forage," which as indicated in Tables 4.1 and 4.2 furnishes about 6 and 14 percent of the feed units fed to beef and dairy cattle respectively, merits attention. On a dry weight basis, this is about 70 percent silage and 30 percent stover and straw. On a feed unit basis, it is probably about 90 percent silage. The silage is about 80 percent corn, 12 percent sorghum, and 8 percent grass silage and haylage (8).

Use of corn silage is increasing rapidly, having doubled in tonnage between 1950 and 1965. This rate of increase will probably continue or accelerate. The rapid increase in use of corn silage has accompanied a decline in the use of pasture by dairy cattle. The percentage of feed units provided to milk cows by pasture has declined by about 28 percent since 1950 with about two-thirds of this decline coming since 1960. Part of this has been offset by green chop and silage and part by higher concentrate feeding.

Some may question the inclusion of corn silage as a forage because of its relatively high grain content in silage as it is made today. How-

TABLE 4.5. *Populations of dairy and beef cattle and consumption of concentrates per head*

Type	1949	1959	1966	1949	1959	1966
		(000 head)			*(lb concentrate/head)*	
Milk cows	23,853	19,527	15,183	1558	2495	3279
Other dairy cattle	12,572	11,073	7,671	458	628	699
Cattle on feed*	6,467	13,085	22,001	2339	2191	2455
Other beef cattle	35,071	58,101	64,114	185	195	220

Source: Data for 1949 and 1959 from Allen and Hodges, 1966 (2); 1966 data from Allen and Devers, 1967 (1).
* Number fed during year.

ever, it has been so considered in agricultural statistics and this procedure has been followed for this discussion. Even if the grain content of silage were removed from this category, it would change the overall percentages by relatively small amounts. The largest change would be for dairy cattle; the change for other ruminants would be insignificant.

Consideration of the structure of cattle populations is helpful in understanding the interrelationships among the various sources of feed units. Populations of dairy and beef animals and concentrate consumption per head are given in Table 4.5 (1, 2). The decline in numbers of dairy cattle and the sharp increase in numbers of beef cattle are readily apparent. The large increase in concentrate consumption per dairy cow and a rather steady level of feeding concentrates to beef cattle are also apparent from 1949 to 1966.

The increase in tonnages of feed grains fed to cattle can be accounted for to a considerable extent by the much heavier per head level of grain feeding to dairy cattle and a substantially increased number of beef cattle on feed. At the same time, consumption of forages also increased and the percentage of feed units provided by forages for cattle (about 75 percent) has remained essentially the same since 1960 (1).

Byerly (4) questioned these statistics and raised doubt that cattle have consumed the amount of forages indicated above. It is of interest in this connection to compare numbers of beef cattle on feed with "other beef cattle," the largest consumers of forage. The number of beef cattle has slightly more than doubled since 1950. The ratio of other beef cattle to beef cattle on feed has decreased from 6:1 to 3:1. This probably reflects some gains in reproductive efficiency, but mostly cattle are going on feed at earlier ages. Other beef cattle, which comprise 75 percent of the beef cattle population, are fed few concentrates. Forages provide about 92 percent of their feed units. About 70–75 percent of this is derived from pasture and 12–15 percent from hay (8). Considering prices for beef calves and feeders, feeding of considerable quantities of concentrates to other beef cattle is economically prohibitive. Therefore, the increase in numbers of these cattle (some 30 million since 1950) must be dependent on increased forage production, and it would appear that the dependence of beef cattle on forages has not decreased sharply. Surely we could have expected modest gains in forage productivity as well as in other crops, during the 1960s and 1970s to provide the feed needed for this increase.

CURRENT AND FUTURE TRENDS

THERE HAVE BEEN much speculation and discussion regarding world population, food supplies, and the role of livestock in the future agriculture of this and other countries. A rather frequently expressed belief is that as the human population explosion intensifies and greater demands are placed on cereal grains for direct human consumption, livestock production will be impractical in the competition with humans for food resources. On the other hand, Byerly (4) states that human population pressure is not likely to eliminate livestock as producers of human food; their scavenger role is too important. Moore et al. (10) emphasize the important role livestock can play in producing animal proteins from feeds not utilizable by humans. They conclude that the role of forages in livestock feeding can be expected to increase rather than decrease. It is not the purpose here to enter into the debate on the future role of livestock as contrasted to direct consumption of cereals as food sources for humans. It appears certain that in the United States, beef and dairy products will be important food items for many years to come. But it seems worthwhile to examine current trends in production of these food items as well as possible future trends.

The amount of concentrates fed to cattle, especially dairy cattle, has increased markedly during the 1960s and 1970s. This has been due to increased technology applied in production of feed grains, which has resulted in substantial yield increases and, consequently, increased availability and lowered prices as well as shortage and increased costs of farm labor. These effects, along with the failure of farm products to advance at rates comparable with purchased items, have intensified the trend to mechanization and automation of farming (including feeding operations) and have increased the size of production units. The trend to fewer and larger production units and increased mechanization is expected to continue at perhaps an accelerated rate. For example, Heady and Ball (7) state that as few as 500,000 farms could readily produce the nation's required output, while Ruttan (12) suggests the number may well decline to 50,000 to 100,000.

Various projections have been made regarding future demands for agricultural production. The report of a joint USDA–state agricultural experiment station long-range study committee (14) has estimated that we will require productivity increases of about one-third by 1980 to meet domestic and export requirements. It has been stated that by the year 2000 an agricultural production double that of today will be necessary. Accompanying this projected increase in demand is a simultaneous per capita increase in beef consumption and a decrease in per capita consumption of milk. To meet these overall goals, production must increase about 2.2 percent annually to meet the 1980 projection and increase an average of about 3 percent annually to double by 2000.

The long-range study committee reported that total agricultural productivity increased about 1.9 percent annually from 1950 to 1973. On the other hand, Daly (5) reports crop output per acre increased 41 percent from 1949–51 to 1964–65, a rate of almost 3 percent a year; and he projects similar increases for the future. The discrepancy in these two estimates lies in the fact that one deals with total production and the other with output per acre. A considerable amount of land in production in 1950 has since been retired in the conservation reserve. Undoubtedly, it is less productive than that remaining in production.

Let us assume that by 1980 we increase beef cattle production one-third and that by 2000 we double it. From Table 4.5 we calculate we would have about 29 million cattle on feed and some 85 million other beef cattle in 1980. Corresponding populations for 2000 would be about 44 million and 125 million respectively. This assumes the present ratios of other beef cattle to cattle on feed. Research to increase reproductive efficiency and feed utilization efficiency could reduce this ratio somewhat, but probably not much in this span of time. To double our beef production, therefore, we must increase our beef cattle numbers by about 86 million head. Other beef cattle receive only about 0.5 pound of concentrate per day on the average and are dependent on forages for about 92 percent of their feed units. With the price structure of today or the foreseeable future, it is not economically feasible to feed concentrates in significant amounts to these animals. Therefore, by 1980 we must provide forages for an additional 21 million head of other beef cattle, and by 2000 we must have forage production capable of supporting some 65 million more other beef cattle.

Doubling productivity of hogs, poultry, cattle, and beef cattle on feed will essentially require doubling the production of feed grains because most of this production is now consumed by these livestock. Gains in livestock-production or feeding efficiency would decrease feed-grain requirements accordingly. Most feed-grain production employs rather high levels of technology including superior varieties, heavy fertilization, herbicides, insecticides, etc. Subsequent increases in technological input are not likely to yield production outputs to the same degree as in the 1960s and 1970s, but new developments such as hybrid barley and wheat and shorter strawed varieties capable of utilizing higher fertilizer inputs will contribute to higher yields, as will still further genetic and cultural inputs to corn. However, it is interesting to note the research reported by Pendleton et al. (11) in Illinois. They found that light became the primary limiting ecological factor when corn was grown under highly productive conditions. Maximum yield in their experiment under very high fertility and irrigation with a high-yielding single-cross hybrid was 23,710 kg/ha (377 bu/acre) with light enrichment. Under normal light, yield was only 55 percent as great. Thus as production inputs increase and yields of corn reach the 11,000 to 12,500 kg/ha (175–200 bu/acre) levels which some farmers are now achieving, further inputs are likely to produce lower returns because of light limitations. However, the return to production of conservation reserve acreages in addition to greater productivity gains probably will provide feed-grain production adequate to meet domestic and export needs until 2000, with livestock feeding levels and feed-grain prices near present levels. If population demands for cereals for direct consumption increase more than anticipated or if the level of feed-grain production increase lags behind demand, feed grains could easily increase in price to the point where economic restrictions would ensue on their use as livestock feeds. In such an event, their use for beef cattle probably would be restricted first because beef cattle are the least efficient of all livestock in converting feed grains to edible human food. The reliance of the beef cattle industry on forages would become even greater.

It is not likely that substantial advances in real cost in the production of beef can be tolerated in the system because of consumer resistance to retail price increases and the prospect of loss of market to imported beef from Argentina and New Zealand (where beef is produced almost entirely on forages) and loss of market to meat substitutes. Neither is it likely that substantially increased production costs can be absorbed by cattle producers.

Forages are basic to cattle production in this country. Very large projected demands for beef production in addition to probable competition for grain crops between animals and the human population of the United States and possibly the world could result in even greater dependence of cattle production on forages. There is no likely substitute for forages as the basic feed in growing replacement dairy animals or for furnishing about half the feed units for high-producing milk cows. It would seem that beef cow-calf operations must be largely based on forages plus minimal concentrate levels for a long time to come. Harvested forages may become considerably more important, accompanying higher production inputs. If human competition for grain crops results in elevated prices, forages could well provide the most economical feed units for milk production and beef animals might remain on high-quality forage for a longer time prior to a shortened fattening period.

POTENTIAL FOR INCREASED PRODUCTION

IT WOULD APPEAR that we have the capability to support substantially increased numbers of cattle without undue difficulty. Several important developments relate to this potential. Technological inputs in production of perennial forages have been very low, even though ample research knowledge is available to demonstrate that application of improved technology is economically profitable. We have improved varieties, and we know how to grow and fertilize them. Yet much forage land grows unimproved varieties that are not fertilized and often are poorly managed. However, farmers are beginning to realize the potential of applying advanced technology to forage crops production. As an example, consider the 10-ton alfalfa club organized in Indiana. Rapid advances in application of new technology to forage crops are expected in the 1970s and 1980s. However, more research and increased educational efforts are required to provide additional technology in the form of still better varieties and the ability to prescribe the most profitable programs for and utilization of these varieties in production of pasture, hay, or silage. Also, of special importance, we must demonstrate this profitability to farmers.

Important gains have been made in harvesting, processing, and storing forages; we are harvesting earlier and getting improved quality and yield, but we still have a long way to go. Von Bargen (17) has stated that over 50 percent of the production cost of alfalfa hay is the result of operations beginning with cutting. Hall (6) estimates that 28 percent of total production of the hay crop is lost between the cutting and feeding operations. These factors seriously reduce profitability of forage crops and make them much less competitive. More research is required to develop new concepts in harvesting, handling, and preserving forage and in treatment of high-fiber forages to increase digestibility. An intensive effort bringing production, harvesting, and utilization of forage into a single integrated program is needed.

More attention will be given to animal performance in breeding and management research on forages. Burton et al. (3) have demonstrated the effectiveness of selection for dry matter digestibility in bermudagrass, *Cynodon dactylon;* and a new variety selected on this basis has been developed and released. Average daily gain increases of 30 percent for beef steers were predicted for the variety. Burton et al. concluded that the nylon bag method, or other *in vitro* technique, could be used profitably to screen genotypes for quality. The chemical techniques of Van Soest may also be valuable in this regard (16). Particularly interesting were the findings that correlations between conventional estimates of quality (percent dry matter, protein content, leaf percent, and crude fiber) and dry matter digestibility were too low to be of value as indices of quality. Animal performance is highly dependent upon intake, which fortunately is positively and significantly correlated with digestibility. We will most certainly direct more efforts toward selecting varieties on the basis of dry matter digestibility and pay less attention to protein content, leaf percent, color, etc. It may be necessary to evaluate many widely divergent genotypes in the search for types with highest dry matter digestibility. Julén and Lager (9) in Sweden have reported heritable differences in digestibility in orchardgrass, *Dactylis glomerata.* Management practices also should be evaluated for their effects on dry matter digestibility.

Protein content has been very important in determining quality of forage for lactating and growing animals. However, the rapid increase in availability and use of nonprotein nitrogen, principally urea, in feeds for ruminants appears to render protein content of forage less important. Future developments are likely to increase effectiveness of urea and also decrease price. This may result in sharp changes in feeding practices for cattle. Agronomists should examine this critically. In this connection, Moore et al. (10) present an interesting discussion of the potential of using nonprotein nitrogen and high-fiber roughages as feeds for ruminants.

Beef animals outnumber dairy animals by 4:1, and the ratio is widening. By 1980, based on estimates previously discussed, the ratio will probably be something iike 6:1, and by 2000 it might be 8:1 or more. More of our future forage research should accordingly be oriented toward beef cattle production. This is not to say that research on production of high-quality forage for dairy cattle should be neglected, but forage for the beef cow-calf operation may be quite different than that for high-producing milk cows.

The beef cattle population is moving eastward. In 1950 about two-thirds of the beef cattle not on feed were in the Great Plains states, mountain states, and the Pacific Coast states. In 1973 only about half were in those regions (2). The greatest percentage increase between 1950 and 1960 occurred in the Corn Belt and Lake states. Surprisingly, the South and Southeast maintained about a constant percentage, roughly 18 percent.

Future increases in beef cattle numbers probably will be concentrated in the Corn Belt, Lake states, and southern and southeastern states. Western range carrying capacity probably will increase with improved practices, but precipitation limitations will tend to restrict growth in beef cattle numbers in the West.

The potential for increasing numbers in the more humid areas of the country is much greater. Release of grazing lands from dairy cattle use, improved yields of forage crops, and availability of large quantities of roughage as a by-product of grain production provide adequate room for expansion in beef cattle numbers in the Corn Belt and Lake states. The potential for beef cattle production in the South and Southeast is also impressive. Upgrading of livestock quality and adaptability is occurring rapidly. Long grazing seasons with high yields of good forage from improved forage varieties and cultural practices seem to ensure adequate and economical production of beef cattle in that area.

CHAPTER 4

SUMMARY

THE ROLE of forages in cattle production is much greater than generally recognized. Their importance is not likely to diminish in the future and could become relatively much greater. Forages have economic value, chiefly as cattle feed. Therefore, the nature of forage production and research should be closely keyed to cattle production practices. The probable trends in the structure of cattle populations and the anticipated requirements for forages in cattle production are of special significance to agricultural administrators as well as forage research and extension personnel. One criterion for determining the significance of forage research is the probability that the results will be of value when the study is completed. Because of the probable increases in cattle numbers by the year 2000, it is especially important that significant forage research receive adequate support.

REFERENCES

1. Allen, George C., and Margaret Devers. 1967. Livestock feed relationships, suppl. USDA Stat. Bull. 337, Tables 27 and 28.

2. Allen, George C., and Earl F. Hodges. 1966. Feed consumed by various classes of livestock. USDA Stat. Bull. 379, Table 3.

3. Burton, Glenn W., Richard H. Hart, and R. S. Lowery. 1967. Improving forage quality in bermudagrass by breeding. *Crop Sci.* 7:329–32.

4. Byerly, T. C. 1966. The role of livestock in food production. *J. Anim. Sci.* 25:552–66.

5. Daly, Rex F. 1967. Agriculture: Projected demand, output, and resource structure, pp. 82–119. CAED Rept. 29. Iowa State Univ., Ames.

6. Hall, C. W. 1957. Drying Farm Crops. Edwards, Ann Arbor, Mich.

7. Heady, Earl O., and A. Gordon Ball. 1967. Research implications of farm firm changes needed in response to wage rates and increased demands, pp. 160–79. CAED Rept. 29. Iowa State Univ., Ames.

8. Hodges, Earl F. 1964. Consumption of feed by livestock 1940–1959 (updated). USDA Prod. Res. Rept. 79.

9. Julén, Gosta, and Agneta Lager. 1966. Use of the *in vitro* digestibility test in plant breeding, pp. 652–57. Proc. 10th Int. Grasslands Congr.

10. Moore, L. A., P. A. Putnam, and N. D. Bayley. 1967. Ruminant livestock, their role in the world protein deficit. *Agr. Sci. Rev.*, Vol. 5, No. 2.

11. Pendleton, J. W., D. B. Engle, and D. B. Peters. 1967. Response of *Zea mays* L. to a "light rich" field environment. *Agron. J.* 59:395–97.

12. Ruttan, Vernon W. 1966. Agricultural policy in an affluent society. *J. Farm Econ.* 45:1100–20.

13. USDA. August 1966. Farm income. FIS 203 Suppl., Table 1.

14. USDA. 1966. A national program of research for agriculture.

15. USDA. August 1967. Feed situation. FdS 220, Table 2.

16. Van Soest, P. J. 1967. Development of a comprehensive system of feed analyses and its application to forages. *J. Anim. Sci.* 26:119–28.

17. Von Bargen, Kenneth. 1966. Systems Analysis in Hay Harvesting. Trans. Am. Soc. Agr. Engrs. 9 (6):768–76.

5

PLACE OF FORAGES IN ANIMAL PRODUCTION—
NOW AND IN THE FUTURE

R. E. HODGSON

THE MOST ABUNDANT FORAGE is perennial grass. Senator
John J. Ingles of Kansas said in 1872, "Grass is the forgiveness of
nature—her constant benediction. . . . The primary and universal
food is grass." He was eulogizing grass (forage) for its contribution
to nature and the welfare of man. He might well have been eulogizing the animal also, for the animal eats the grass and nourishes man.

Forage covers the open spaces of the earth and complements the
forests. It graces our homesteads and serves to protect and maintain
our soil and water. It contributes greatly to livestock feeding and
thus indirectly to our generous, varied, and abundant food supply.
However, except for the ruminant sector of the livestock industry, it
would have little market value.

IMPORTANCE OF FORAGES TO LIVESTOCK PRODUCERS

THE LIVESTOCK PRODUCER's dependence upon forage is limited to a
large degree to feeding ruminant and ruminantlike animals—cattle,
sheep, goats, and horses—because anatomically these animals have
rumens or large cecums that permit them to consume and hold large
quantities of coarse materials as feed. These organs are virtually fermentation vats where microbes (bacteria and protozoa) break down
cellulose and other fibrous material by chemical and enzymatic processes, thus releasing and changing nutrients and otherwise making
them available for absorption and use by the host animal. We are

Based on an address presented at the American Forage and Grassland Council Research-Industry Conference, Chicago, Illinois, January 1968.

57

greatly indebted to these microscopic organisms without which their hosts could not produce the abundant supply of high-quality meat and milk we enjoy.

The 1964 census (9) tells us that we have a total land area in the 48 contiguous states of 1902 million acres. Of this, 1106 million acres was land in farms and 796 million acres not in farms. Thirty-seven percent, or 294 million acres, of the land not in farms was used as grazing land. Of the land in farms 702 million acres, or 63 percent, was used as grassland (permanent), woodland, and cropland pasture, and for the production of hay and corn silage. Therefore, of our total acreage 52 percent is used for growing forage of some kind.

It is difficult to estimate the monetary value of forages as used by our ruminant livestock population. Hodgson (2) arrived at a figure of approximately $8.0 billion for 1965. (See Chapter 4.)

CONTRIBUTION OF FORAGE TO FEED SUPPLY

THE BEST INFORMATION on the contribution of the forage resource to the total feed consumed by livestock is provided by the reports on livestock-feed relationships issued by the USDA (Table 5.1) (6). For all livestock the feed unit equivalent consumed from roughage, hay, and other harvested forage (including silages and pasture) amounted to 202.6 million tons or 53.8 percent of the total consumed. For all cattle, sheep, and goats the total for forage accounted for nearly three-fourths the total feed units consumed. The percentage for dairy cattle was 67.4 percent; for beef cattle, 76.5 percent; for sheep and goats, 90.4 percent; for hogs, 4.2 percent; and for horses and mules, 78.4 percent. The contribution of forage feed units to poultry and other livestock feeding is negligible.

Sheep and goats get most of their feed from pasture and range. Dairy cattle and other than milk cows obtain a higher percentage of their feed from harvested forage and especially pasture. In contrast to a decrease of about 35 percent in the number of dairy stock since 1950, the beef cow population has about doubled. Beef cattle fed out in feedlots have about doubled as well. When on feed, they consume much less forage than they do during the growing period. However, some three-fourths of the beef cattle receive the major part of their feed from forage. This must continue in order to keep the cost low for maintaining breeding herds and raising young, growing cattle.

Despite their great quantity and value, forages have been de-

TABLE 5.1. *Consumption of feed in feed units by different classes of livestock and percentage contribution of different kinds of feed,* 1965

Livestock class	Concentrates		Hay		Other harvested		Roughage				All feed
							Pasture		All		
	Total	Percent	Total	Percent	Total	Percent	Total	Percent	Total	Percent	
	(000 tons)		*(000 tons)*		*(000 tons)*		*(000 tons)*		*(000 tons)*		*(000 tons)*
All livestock	173.8	46.2	48.2	12.8	22.8	6.1	131.6	34.9	202.6	53.8	376.4
Cattle and sheep	69.5	26.4	46.0	17.5	22.1	8.4	125.2	47.6	193.3	73.6	262.8
Dairy cattle	30.0	32.6	26.8	29.0	12.7	13.8	22.7	24.6	62.2	67.4	92.2
Beef cattle	38.3	24.2	18.6	12.4	9.0	5.7	92.4	58.4	120.0	76.5	158.3
Sheep and goats	1.2	9.6	0.6	5.0	0.4	3.2	10.1	82.2	11.1	90.4	12.3
Hogs	54.3	95.8	...	...	...	...	2.4	4.2	2.4	4.2	56.7
Horses and mules	1.5	21.8	1.5	21.7	0.6	10.2	3.2	46.4	44.9	78.4	46.4

Source: USDA, 1966 (6).

clining in use as feed for livestock. This is particularly true in the case of dairy cattle. In 1955 forages contributed 56.4 percent to the total feed unit supply as compared to 53.3 percent in 1965. A number of reasons account for this: (1) the relatively poor quality of hay; (2) the relatively low yield of pasture and perennial-type harvested forage; (3) the risks and difficulty of harvesting and storage; and (4) the abundance, high quality, and low cost of feed grains. The latter, no doubt, is the principal reason. One must ask the question, "Will this favorable feed-grain situation continue to prevail?" Or conversely, "Will forages continue this trend toward making a smaller contribution to the feed supply?" Probably, with dairy cattle at least, it might continue to decrease unless a concerted effort is made to improve our forages and reduce their costs.

THE PLACE OF FORAGES IN THE FUTURE

A PROJECTION of forage use in animal production can be made by looking at future needs for livestock products and assuming that forages will not change in importance. The USDA Economic Research Service prediction of the 1980 production of animal foods is given in Table 5.2 (8).

It is not easy to translate these figures into requirements for different feeds for the 1980 production. Table 5.3 provides some estimates (6, 7). First it was necessary to translate carcass weights to comparable liveweights. This was done by the use of Tables 466, 468, 502, and 505 in the USDA 1967 agricultural statistics. Then Table 539 was used to obtain the feed units consumed for the production of given units of beef, sheep and lamb, and milk. The 1964 data on all forage feed units consumed were used as the base. The feed units needed for 1980 were estimated by applying the feed units consumed per pound of liveweight production or milk production in 1964 to

TABLE 5.2. *Predicted production of cattle and sheep products*

Product	Base 1964	1980	Increase	Percent
		(mil lb)		
Beef and veal (carcass wt.)	19,407	27,000	7,593	39.1
Lamb and mutton (carcass wt.)	716	818	102	14.3
Milk	126,598	145,289	18,691	14.8

Source: USDA, 1967 (8).

TABLE 5.3. *Estimate of production of animal products (fresh weight), total feed units required from all feeds and from forage*

Item	Cattle and calves	Lamb and mutton	Milk
1. Estimated liveweight production (000 tons)			
1964	16,339	745	63,299
1980	23,200	852	72,644
Increase	6,861	107	9,345
2. Estimated feed units required for production (000 tons)			
1964	173,528	12,478	71,477
1980	232,000	13,463	82,815
Increase	58,472	985	11,338
3. Estimated feed units from forage			
1964	138,963	11,455	45,568
1980	185,600	12,251	52,575
Increase	46,637	896	7,005

Sources: USDA, 1966, 1967 (6, 7).

calculate the tonnage of feed units needed for the 1980 production. The percentage contribution that forage made to the total in 1964 was used to estimate the forage feed units needed in 1980. These data are given in part 3 of Table 5.3.

It is not claimed that this is an accurate prediction for feed nutrients needed from forage by 1980, but it is believed to be a reasonable estimate. On this basis, an estimate of the needed increase of forage feed units for beef cattle amounts to 46.6 million tons; for milk cows, 7.0 million tons; and for sheep, 896,000 tons; or a total of 54.5 million tons. This would indicate a 28 percent increase in feed units by 1980 when compared with figures for 1964. This may be an overestimate of the forage requirement, particularly if forages play a lesser role in the feed supply than they do at present. It would appear that we should anticipate the need for about a one-third increase or 50 million tons of forage by 1980. This should be a reasonable objective.

Achievement of Future Supply. It is unlikely that much more land will or can be devoted to forage crop production. Indeed there may be less as pressures for land from different sectors increase. It appears that expanding needs for feed units from forage must come from greater yields and from increased efficiency in the use of the herbage that is grown. There are great opportunities for improvement in both as well as an opportunity to make greater use of crop refuse in combination with nonprotein nitrogen and through new chemical and physical processes.

In our concern about population pressure, how to feed ourselves and the world's less fortunate, and agricultural production and planning, we hear little about the importance of the great forage resource. As the pressure for food becomes greater and the use of grains for direct human food and the value of these grains increase, their cost as livestock feed may reach the point where even greater dependence must be placed on forages if livestock products are not to be priced off our tables. It would be desirable for the livestockman, regardless of the pressures on the feed grains, to get more out of the forages he grows.

FORAGE COSTS AND YIELDS

YEARS AGO it was easy to state that forages were the least expensive source of nutrients for lactating cows. This is much more difficult today because (1) the acre yields of feed-grain crops have increased more than yields of forage crops; (2) the production, harvesting, and handling of grains have been mechanized and simplified to a much greater degree; (3) grains are easier to handle and contain more feed nutrients; and (4) grain quality has increased and is standardized.

A good illustration of this relative shift in cost of nutrients in forage and corn grain over time has been assembled by Gordon (Table 5.4) (1). In 1955, for example, with hay valued at $17.60 per ton and corn at $1.30 per bushel, the total digestible nutrient (TDN) cost was 70 percent greater for corn than for hay. In 1965, corn cost

TABLE 5.4. *Relative price of feed nutrients from hay and grain crops in Wisconsin*

Feed	Source	TDN costs, 1955			TDN costs, 1965		
		Price/ unit	Cost per ton	Rel- ative	Price/ unit	Cost per ton	Rel- ative
Hay, all	Wis.	$17.60	$34	$100	$24.50	$48	$100
Wheat	Wis.	1.85	77	226	1.38	58	120
Corn	Wis.	1.30	58	170	1.10	49	102
Oats	Wis.	0.60	54	158	0.67	60	125
Barley	Wis.	1.08	58	170	1.02	55	114
Soybeans	Wis.	2.15	81	238	2.35	88	183
Sorghum	Nebr.	1.02	45	132	0.91	40	83

Source: Gordon (1).

TABLE 5.5. *Changes in yield per acre of certain feed crops*

Crop	Unit	1954	1965	Percent increase
Corn grain	bu	37.1	73.1	100
Sorghum grain	bu	19.0	50.1	164
Wheat	bu	18.1	26.9	43
Oats	bu	35.6	50.2	41
Barley	bu	28.5	43.5	53
Soybeans	bu	20.1	24.4	21
Corn silage	ton	7.5	10.6	41
All hay	ton	1.43	1.79	25
Alfalfa, alfalfa mix	ton	2.10	2.48	18
Clover, timothy mix	ton	1.43	1.53	7

Sources: USDA, 1955, 1966 (3, 5).

was only 2 percent greater compared to hay. Much of this difference is due to the great increase in the yield of corn.

The operators of large feeding units usually purchase much of their feed and are very sensitive to relative costs in the feed market. They are not the least sentimental about forages and will turn to whatever feed sources of energy and protein are cheapest. If the producer of forages wishes to participate in this market, he must meet the competition. It is likely that feed evaluation will be based on net energy rather than TDN in the future. If this occurs, the position of forages may be even less favorable.

The per acre yield of hay has not increased as much as that of feed-grain crops. Between 1954 and 1965 the yield of all hay increased 25 percent as compared to corn, 100 percent; sorghum grain, 164 percent; and barley, 53 percent. Other comparisons are shown in Table 5.5 (3, 5).

It is doubtful that the per acre yields of pastures have increased any more than for hay. The only bright spot in the forage picture is corn silage which increased 41 percent. Many individual farmers make good money with forages by producing yields much above the average at lower than average costs.

The huge feed-grain and associated seed by-product supplies are recognized as an important source of feed for livestock, including ruminants. Producers should continue to use them to their advantage. These feeds have made a most significant contribution to increases in livestock production and will continue to do so.

The forage supply is largely used on the farms where it is grown as an essential part of the farm management and cropping system. Crops, whether grown for pasturage or harvested forages, should be developed to their full economic potential and be used liberally in feeding the farmer's livestock. They make the major contribution to the feed supply, but this could and should be greater. They should be supplemented by grains and other concentrates to provide the proper nutritive balance and should be fed to the extent that each added increment returns more than it costs.

A concerted, well-directed effort in research, education, development, and application in the production, conditioning, management, and utilization of all available forages is needed to enable them to increase their contribution to agricultural production and to the animal food supply of the future.

We need better species and varieties of forages that are higher yielding and tailor-made to perform under local solar energy, temperature, soil, and water conditions. They should be developed with the objective of providing livestock nutritional needs, not just the production of a quantity of dry matter. Attention needs to be directed toward growth habits in relation to feed needs throughout the year and the ability to carry nutrients through the conditioning procedure to the manger. Particular attention needs to be given to developing palatable crops that have a high intake potential when fed.

Soil, water, and nutrient supply management need to be geared so that the land affords the proper environment for improved varieties of forage plants to combine the energy from the sun and the nutrients from the soil to metabolize and store in the tissue nutrients that are palatable and readily available to the animal.

IMPROVEMENT OF FORAGE PRODUCTION AND USE

IMPROVEMENT in harvesting techniques would very significantly increase both the acceptability of the forage to livestock and the amount of protein and other nutrients that actually reaches the animal. This is particularly true in the humid areas where climatic conditions favoring forage production also make crops difficult to harvest. Experiments conducted at the USDA Research Center at Beltsville in the 1950s have effectively demonstrated the great losses that occur in the nutrients of forages during harvesting and storage. The potential saving from reducing these losses is illustrated in Table 5.6 (4).

TABLE 5.6. *Effect of reducing harvesting loss on hay costs*

Item	Harvesting and storage loss (%)			
	30	20	10	0
Net yield/acre (tons)	2.30*	2.64	2.98	3.30
Cost/ton	$24.00*	$20.83	$18.46	$16.67

* Yield and farm value of hay, north central states (4).

The preservation of the hay crop as grass silage is illustrative. Losses can be cut in half by using the silage procedure instead of harvesting for hay. Surely other perhaps more efficient economical methods can come from research development applicable to the individual farm. New machinery and alternative techniques have been developed and are in efficient use. These have perhaps done more to reduce the labor and ease the hard work of harvesting than to maintain and improve quality or to reduce nutrient losses. The farm equipment industry and the researcher have large roles to play here, more than the contributions each has made in the past.

Much of the forage supply is grazed where the animal does the harvesting job. The management of grazing as well as the management of the pasture or range area involved needs improvement. Brush and weed control and grassland restoration on much of our range and permanent woodland pastures are examples of this need. Under intensive livestock production such as dairying, it appears questionable that grazing is the single best way to utilize forage. Under usual conditions only 70–80 percent of the available pasture forage is actually utilized. In the modern intensive farming scheme, we must wonder if alternatives to grazing are not needed and have seen trends in this direction. This might be a good way to minimize climatic stress effects on high-producing animals in the South when on pasture.

Many advances can be made in production and conditioning to improve forage acceptability, and much can be done with the animal to increase consumption of forage. This includes possible pretreatment of the forage and changes in the time of feeding and the form of the forage as well as developmental selection of animals that are big forage consumers. This opportunity rests with the animal breeder.

We must have greater understanding and better management of the digestion and metabolism processes of the animal. Forages, except in an early stage of development when they tend to resemble concentrates chemically, are characterized by a high fiber content. This involves special problems in digestion since the lignin of fiber tends to tie up the desirable nutrients and to make them less available. In an attempt to study this and other digestive problems, scientists have learned how to put windows in the rumen. Through such techniques, coupled with biochemical and physiological research on both the animal and the forage material itself, scientists are beginning to gain greater understanding. This approach offers good opportunity to improve the acceptability and utilization of forage by the ruminant.

In addition to the more conventional forages used as livestock feed, huge supplies of low-grade materials such as straws and other by-products are available. In their existing form they have low acceptability and are high in fiber and low in nutrient content. Possible new chemical and physical treatment and supplementation with nonprotein nitrogen and other materials could upgrade these materials as feed. They could be especially useful for maintenance, if not for production, and for rations or supplements. They could also be used for the raising of young beef and dairy stock from the milk-feeding stage to the feedlot stage at low costs.

FORAGE FOR OTHER ANIMALS

ALTHOUGH attention has been directed to the ruminant, we should not overlook the contribution forages can make to other classes of livestock. The resurgence of the horse industry in our affluent society directs attention to forages as a principle source of feed for this species, although some research problems are involved. Also, forages will play a more important role in hog production and, to some extent, for certain classes of poultry.

Rural poverty may be alleviated by the raising of small quantities of goats, rabbits, ducks, geese, and other small species as well as larger animals. Home-produced forage should form an important part of the feed supply for these small livestock enterprises.

Forages have been regarded as a most essential input in the livestock production enterprise. In the USDA animal research organization an important part of the work is devoted to developing information to improve the utilization of this kind of feed by the different classes of livestock; greater efforts are anticipated.

At the U.S. Meat Animal Research Center in Clay Center, Nebr., part of the research program is being directed toward beef cattle and sheep production in the context of utilizing forages to the best advantage and more efficiently producing meat products of higher consumer quality. A need is anticipated for a similar effort in the humid north central region where intensified research on the dairy cow that is directed to reducing the cost of milk production will be undertaken. This added research effort will greatly complement the state-federal cooperative programs now under way.

SUMMARY

THIS REVIEW indicates that forages have been, are, and will continue to be the major source of feed to support our huge livestock industry. There is a highly mutual compatibility between utilizing land in the production of forages and in livestock production. Forage would have little value save for livestock. Livestock production would be difficult and costly without forage feeds. Forages generally have not shown the same level of improvement in production and conditioning as most other crops. To rectify this, a wide spectrum of research interests must concentrate on the contribution of forages to the livestock feed supply and the agricultural economy. The animal husbandman has the conviction that this attention is greatly deserved.

REFERENCES

1. Gordon, C. H. Unpublished data. USDA.
2. Hodgson, H. J. 1966. The importance of forage in livestock production in the United States. Annual meeting, Am. Soc. Agron., Washington, D.C.
3. USDA. 1955. Agricultural statistics.
4. USDA. 1965. Agricultural statistics.
5. USDA. 1966. Agricultural statistics.
6. USDA. 1966. ERS Stat. Bull. 337.
7. USDA. 1967. Agricultural statistics.
8. USDA. 1967. A look ahead for food and agriculture. Agr. Res. Serv. Bull. 18.
9. USDC. 1966. 1964 U.S. census of agriculture. Bureau of Census.

6

PRESERVING GRASSLAND PRODUCTS FOR RUMINANT FEEDING

CHESTER H. GORDON

THE NEED for forage preservation has no place in the ideal concept of a static balance between the nutrient requirement of grazing ruminants and the nutrients produced by the growing forage plants. Such an ideal system obviates the need for preservation and its attendant problems. Although this is conceptually attractive, the idea is predominantly unrealistic because the nutrient requirements of any given population of grazing animals and the productivity of forage plants are dynamic factors resulting in frequent imbalances. Although the absence of preservation may be noted in a number of grassland regions, extended periods of animal:plant imbalance occur. These simplified systems may persist indefinitely and may in fact be most profitable, but they can seldom be defended on the basis of biological efficiency.

The nutrient requirements of grazing animals are increased by advancing pregnancy, growth, and higher levels of milk production. Conversely, requirements are decreased during nonpregnant and nonlactating periods (34). When animals must be protected from severe winter weather, the opportunity of direct grazing is eliminated even if this were possible from an agronomic standpoint. The need for protection from any extremely hot environment may also preclude grazing.

Essentially all grassland areas of this country are characterized by an early spring flush of growth. The period is marked by cool weather, long daylight hours, and ample moisture. Forage produced in this period is not only abundant but also highly nutritious and palatable, whereas forage production following this optimal period is normally reduced in both quantity and quality. For example, it was reported in Maryland that the July rate of total digestible nutrients accumulation in orchardgrass-ladino pastures was roughly one-half the rate observed in May and twice that observed in September (42). During a six-year trial in Tennessee (3) grazing was interrupted five times because the supply of forage was exhausted despite stocking rotational pastures at less than one cow per acre. This seasonally cyclic growth pattern can be overcome to some extent through a skillful integration of varieties and species in a total forage program. However, seasonality is still a striking characteristic of grassland production.

The need for forage preservation is also based upon human and economic factors. The optimum manipulation of stocking rates and pasture resting periods is highly complicated and demands the frequent attention of an interested and astute manager. Many livestockmen prefer to avoid this biological juggling act and devote their time and attention to other activities. Hoglund (24) reported that a majority of 182 Michigan dairymen polled had reduced their full-season pasture acreage by at least 50 percent during the 1960–65 period and were depending more heavily on preserved crops or annual pastures. The majority of livestock farmers have a certain amount of equipment on hand that is used to store and preserve forage. This investment is frequently necessary for accommodating a winter feeding period; hence feeding from storage during the summer does not increase their capital investment. The continuing trend toward larger livestock feeding operations makes direct grazing of forage crops considerably more difficult from the standpoint of the mechanics of moving animals from the feeding or milking areas to available grazing areas. In fact, milking areas are sometimes several hundred miles from the areas of forage production.

The social and economic pressures for maximum efficiency in forage production and utilization are greatest when forage is produced on land that is also capable of growing crops for direct human use. The wastes of grazing high-yielding pastures have been reported as 4–7 percent (19, 27) and are viewed with alarm. The interest in preservation is proportionately diminished in less productive areas since lower yielding crops are grazed with less waste and are more expensive or impossible to harvest mechanically, and no economically feasible alternative utilization may be available for the more extensive areas.

SELECTING FORAGE MATERIAL FOR PRESERVATION

ONCE A COMMITMENT to preserve forage has been made, selecting the harvesttime is one of the first and most important decisions. Essentially all grassland crops are characterized by gradually declining chemical quality as the plants approach physiological maturity (11, 44, 46). The increase in first harvest yield, generally observed by delaying harvest date, must be recognized as a partially compensating factor. The net seasonal advantage of this increased yield is reduced, however, by leaf loss from lower plant parts and a general reduction in the yield of successive crops. No net seasonal yield advantage was found by Dawson et al. (11) by delaying the harvest of alfalfa beyond the initial bloom stage under Montana conditions.

The generally declining digestibility with advancing date of harvest in any particular location has led to the suggestion that cutting date is a reliable guide for predicting digestibility coefficients without regard to the plant species involved (1). An equation for predicting dry matter digestibility proposed by Reid et al. (39) included a regression of 0.48 units in the digestion coefficient for each day of harvesting delay. This group reported a standard error of estimate of 1.65 percent and found only minor differences among the species of perennial forages. This relationship was verified for timothy by Poulton et al. (37), although Colovos et al. (8) found significant differences in varieties of timothy harvested on the same date.

Prediction of forage digestibility from cutting date apparently works within rather broad limits of error simply because the range of economically important forage plant species within any particular location is not very wide. However, the more serious practitioners of preservation of grassland yield must recognize distinct forage species differences in the levels and rate of decline in digestibility and voluntary intake (10, 33, 46).

METHODS FOR PRESERVING FORAGE AND ROUGHAGE

AFTER A PLANT IS CUT, the chemical degradation of plant material starts almost immediately through continued respiration or through the activity of microorganisms (molds and bacteria) (5). Both types of degradation may be arrested by desiccation, excluding oxygen, acidification, or some combination of these. Preservation methods are selected and evaluated primarily on the basis of their ability to arrest degradation and retain feeding value. While a complete review of all methods of preservation is not within the scope of this chapter, the distinguishing characteristics, limitations, and sample references of the major methods are presented.

Drying. Desiccation methods (natural or artificial drying) depend upon reducing moisture content to the point that enzymatic activities of both plant and microbial origin are arrested. To achieve this type of preservation, the moisture content must be reduced to a level below 20–25 percent depending on the plant species, bulk density, ambient temperature, and the relative humidity of the microatmosphere (32, 35). When forages are stored prior to meeting these requirements, heat is generated and chemical degradation continues until such time as the crop becomes dry enough to arrest them. This continuation period may be for a few hours or for a few months depending upon the moisture content at storage, although peak temperatures are typically reached within two weeks (35). Degradation during a few hours is minor and difficult to identify, whereas degradation over a period of weeks or months is so extensive that feeding value is seriously reduced.

Field drying is by far the simplest method of crop preservation, but it is also subject to the greatest field losses and quality variations (23). The losses of nutrients in alfalfa, for example, were shown to increase with longer field exposure since this is associated with greater rain leaching, leaf shatter, and other mechanical losses (41). It is unfortunate that leaves, which are highest in nutritive value, dry faster than stems and are then very easily shattered during the final stages of stem drying. This situation has prompted several attempts to increase the rates of stem drying by mechanical conditioning.

The major effects and interactions of hay conditioning have been reported by a number of workers (28, 29). Conditioning by crusher, crimper, or flail harvester has reduced field drying time by as much as 30 percent. Uniform crushing is generally more effective than crimping for increasing stem drying rates, but it is also more difficult to attain. During rainy periods, treatments which have allowed faster drying also cause water to be absorbed at a faster rate. Under very poor weather conditions this may amount to a net disadvantage from conditioning. The flail-type forage harvester has given maximum increases in drying rates, but field losses may be increased 25–30 percent above those occurring with crimping or crushing. Field-cured hay is frequently baled one day earlier by use of these conditioning methods, and the advantage is most pronounced in thick-stemmed legume crops.

All systems of preservation are as much an art as a science, but the importance of the skill and judgment of the operator is perhaps greatest in the case of field curing. Selection of cutting time, operation of conditioning machines, and times of raking and baling all represent subjective judgments. As a result, variations in field losses and final hay quality are tremendous. Attempts to consistently harvest field-cured hay with minimum damage usually result in a delayed cutting schedule or a large stand-by force of labor and equipment and reduced seasonal forage yield and quality. Neither of these effects is compatible with efficient grassland farming.

The field drying of hay is so dependent on weather that barn-drying with forced air, with or without heat, has attracted considerable attention in the humid parts of the United States and in northern Europe. The various methods of artificial drying (hot air or cold; complete or partial; for long, chopped, or ground forage) increase the possibility of obtaining a satisfactorily low level of moisture in a relatively short time. The enthusiasm of barn-drying advocates is understandable in view of the excellent appearance and biological advantages of the product. In experiments carried out by Shepherd et al. (42), the preservation of initial alfalfa crop dry matter for feeding averaged 85 and 81 percent for hay dried with and without heat respectively. Comparable preservation of field-cured hay was 63–79 percent depending on weather conditions. Crude protein preservation rates of the systems with and without heat were 79 and 76 percent, while field curing preserved 54–72 percent. The improvement in carotene preservation was even more striking with barn-dried hay having two to ten times the carotene content of field-cured hay.

Favorable results from forced-air drying were also reported by Cole et al. (7). An average of 11 percent more crop was preserved by harvesting at 35–45 percent moisture content and barn-drying with heat. Crude protein content of the hay was consistently improved only by harvesting at the higher moisture level. These workers also noted that improved harvesting efficiency and composition were closely associated with harvesting at a moisture content compatible with leaf retention.

Harvesting hay at progressively higher moisture content generally results in progressively higher quality and greener hay if drying can be completed within a few days. This relationship has led many enthusiasts to overlook the economic implications of barn-drying at higher moisture content. Some concept of the magnitude of the

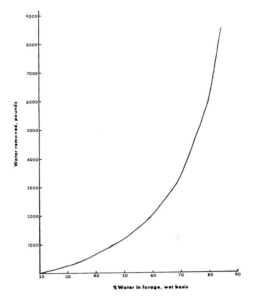

FIG. 6.1. WATER REMOVAL REQUIREMENTS FOR PRODUCING 1 TON OF 80 PERCENT DRY MATTER HAY.

moisture removal at progressively higher moisture content may be obtained from Figure 6.1. We see, for example, that the moisture removal requirement for obtaining 1 ton of hay is more than twice as great at 40 percent moisture (666 lb) as it is at 30 percent moisture (285 lb).

Fuel efficiency for most hot air dryers makes a gloomy picture even worse. Calculations based on Wysong's report (56) show a maximum water evaporation rate of about 21 pounds per gallon of fuel oil burned, although theoretically one gallon of fuel oil should evaporate about 160 pounds of water. Harvesting when moisture content is high may result in harvest delays while waiting for drying to be completed.

Although the increased hay yields, improved chemical analyses, and greater feeding value potential of heat-dried hay are generally accepted, no widespread adoption of the practice has occurred in the United States. In fact, several machinery manufacturers who previously supplied wagon drying systems no longer do so. The possibility of a future major technological change making such on-farm drying economically feasible should not be overlooked however. Much greater efficiency is attained in large commercial dehydrating plants.

However, these installations, with capacities of several thousand tons of crop annually, are not often available for preserving the surplus crops of individual grassland farms.

Ensiling. The many difficulties of drying in humid areas have in no small way contributed to the growing popularity of ensiling, which depends basically on the attainment and maintenance of an oxygen-free condition. The attainment of a specific moisture level, although important, is much less critical for silage than for hay. The amount of oxygen initially present when silage is stored allows the oxidative degradation to continue for at least a short period of time. This activity is very wasteful from an energy standpoint since the primary products are carbon dioxide and heat. Fortunately, the initially trapped oxygen is utilized within a period of one to four hours (26) after which less wasteful anaerobic changes predominate. If, however, an anaerobic condition is not established and the oxidation activities continue, very high storage losses as well as moldy, rotten, and extremely hot silage are observed.

The extent of continuing oxidation is a function of the porosity of the silo seal and porosity of the silage mass itself as well as the chemical composition of the silage (25, 26). Some oxidation and heating is probably unavoidable under farm conditions, but in the United States it should be regarded as primarily detrimental and all reasonable steps should be taken to minimize this period and type of activity.

After anaerobic conditions have been established, the extent and type of further activity depend on the composition of the crops and the microorganisms present. In general, crops with a high level of carbohydrates (particularly water soluble ones) produce high-quality silage with rather low fermentation losses. It is also true that poor fermentation patterns may usually be improved by the addition of carbohydrate materials. However, the different water soluble carbohydrates (i.e., glucose, fructose, or pentose) as well as the specific organic acids present in plant material (i.e., malic, citric, glyceric) have specific influences on the fermentation route (52). It has also been demonstrated that the most active species of bacteria, in fact the particular strain of lactic acid bacteria, will influence the ratio of acids present in final silage as well as the fermentation losses (52).

The suppression of changes in nitrogen compounds is of equal importance in controlling carbohydrate changes. Proteolysis, or simplification of the true protein, begins as soon as a forage crop is cut.

These changes apparently continue in the field or silo until a high enough dry matter or acidity content is attained. The importance of plant enzymes to these changes was demonstrated by the observation that significant quantities of ammonia were formed in bacteria-free silages (36). The major proteolytic changes are brought about by certain bacteria under usual silo conditions. The activity of these organisms is responsible for the formation of various amines, ammonia, and branched-chain fatty acids. These bacteria are also capable of the secondary fermentation of lactic acid to butyric acid and carbon dioxide (26, 52).

Natural Fermentation. It is important to realize that under some crop and storage conditions spontaneous fermentation is extremely efficient, silage is well preserved, and the product is highly desirable. If this fortunate situation exists, it is highly unlikely that any additive or special procedural arrangement will provide any improvement. Conversely, to the extent that these ideal conditions do not exist, it will be relatively easy to demonstrate an improvement. It is not surprising, therefore, that no general agreement exists in the literature on the effectiveness, desirability, or necessity of any particular silage treatment. For this reason some of the more prominent silage treatments will be described in terms of their possible corrective action. This is not to imply, however, that an improvement will always be observed from the use of the technique in question.

Wilted Silages. The reduction of crop moisture content through field wilting has under some circumstances been shown to have beneficial effects on the chemical and feeding characteristics of the resulting silage (16, 21, 41). Under other circumstances, however, no significant improvement has been noted (6). Much of the beneficial effect can be explained on the basis of the sensitivity of the *Clostridium* bacteria to high osmotic pressure. Therefore, as moisture content is reduced through wilting, the concentration of nutrients in the solution and the osmotic pressure of the plant solution increase. The clostridia, which are responsible for much of the degradative changes in silage, are particularly inhibited by this increased osmotic pressure (53). Gouet et al. (21) reported that prewilting of the crop resulted in a slower but continuous production of lactic acid throughout the preservation period of 98 days. The formation of butyric acid and ammonia were remarkably inhibited by this prewilting. These observations may be explained as reduced clostridial activity. Some rather typical effects on the chemical composition of the silage were reported by Gordon et al. (16, 18) as lower levels of ammoniacal nitrogen, total acid, acetic acid, butyric acid, and propionic acid. The content of lactic acid was slightly increased and the proportion of lactic acid greatly increased by lowering moisture content from about 80 percent to the 40–50 percent range. The major part of these changes, however, is attained by reducing the moisture content to about 65 percent (18, 21).

The benefits of wilting involve some biological and monetary costs. Field losses prior to actual harvest can range from 1 to 25 percent. It appears that most of these field losses are associated with the raking or turning of the crop during the wilting process. When wilting in the windrow without mechanical handling is possible, very low field losses are experienced (20). Some alteration of the chemical constituents of the fresh plant and some energy loss are unavoidable during the wilting process (5, 20).

Additional major problems and losses from wilted forage may occur after placement in the silo. Since the forage mass becomes less dense with increasing levels of prewilting, at some level the infiltration of air may be rapid enough to permit destructive aerobic changes, including oxidation and generation of considerable heat. The point at which this becomes a problem is dependent not only on the moisture content of the crop but on such physical factors as the length of cut, extent of consolidation, and effectiveness of mechanical seals in the silo. Aeration and the attendant heat that is produced have been specifically related to a reduction in the digestibility of protein as well as loss of organic matter (14, 25, 54). Frequently, the total crude protein content of forage is not altered appreciably by heating and the usual forage analysis will not indicate that damage has occurred. Some damage, however, is reflected in an increase in the nitrogen content of the acid detergent fiber fraction (47). Therefore, the expected improvements from prewilting (improved chemical quality, improved animal acceptance, improved odor) are attainable only at the expense of increased attention to field operations and more exacting requirements for the preparation and storage of this wilted forage.

STIMULATION OF LACTIC ACID FERMENTATION

SEVERAL METHODS of silage treatment have the common objective of specifically stimulating the growth of lactic acid organisms. Simple mechanical treatments such as very fine chopping, laceration, crushing, or bruising have a profound stimulatory influence on the production of lactic acid in small experimental silos (12, 15, 57). This has been explained on the basis of early liberation of the contents of cells, resulting in a higher osmotic pressure of the silage juice as well as early activity of lactic acid organisms. Improvements from similar mechanical treatments are not consistently observed in large tower silos (2). Pressure within such structures may liberate plant juices quite effectively with only nominal chopping.

Another common practice for stimulating lactic acid is the addition of carbohydrates in the form of cereal grains or molasses. Although such an approach is quite dependable in terms of improving the chemical quality of otherwise problem silages, it may be distinctly expensive in terms of loss of the concentrates. Direct measurements of the grain dry matter losses under those conditions are not possible by usual techniques; but by making certain assumptions, some interesting efficiency relationships may be developed. The comparisons presented in Table 6.1 are based on the following assumptions:

1. Grain available contained 90 percent dry matter.
2. Forage as stored contained 20 percent dry matter.
3. Dry matter losses are related to the dry matter content of ensiled material (Table 6.1).
4. Grain can be stored in a bin with no loss.

TABLE 6.1. *Estimated efficiencies of storing grain and 1 ton of forage by two systems*

System	Grain available per ton of fresh forage* (lb)					
	50	100	150	200	250	300
System I	*Forage and grain ensiled together*					
Dry matter content of mixture stored (%)	21.7	23.3	24.9	26.4	27.8	29.1
Estimated dry matter losses:						
From combined ration (%)	25	22	19	16	13	10
From forage portion (lb)	100	88	76	64	52	40
From grain portion (lb)	12	22	28	32	33	30
Percent grain in dry matter fed	10	18	25	31	36	40
System II	*Forage ensiled, grain binned*					
Estimated dry matter losses:						
From forage portion (%)	26	26	26	26	26	26
From forage portion (lb)	104	104	104	104	104	104
Percent grain in dry matter fed	13	31	31	38	43	48
Comparative advantages of System II						
Forage preservation (lb)	— 4	—16	—28	—40	—52	—64
Grain preservation (lb)	+12	+22	+28	+32	+33	+30

*Grain 90% dry matter, forage 20% dry matter as stored.

These relationships appear to be of economic importance. For example, if one chooses to follow System II at 100 pounds of grain per ton of forage he will have 16 pounds less forage but some 22 pounds more grain; and when these materials are combined for feeding he will have a mixture that is 23 percent instead of 18 percent grain (an improvement of 28 percent).

Grain added to wilted forage is not expected to lose dry matter as extensively in the silo. On the other hand, from the standpoint of fermentation control there is little reason to add grain to silage with lower moisture.

Inhibited Fermentation. Use of bacteriostatic or bactericidal materials represents still another approach toward the control of fermentation. Sodium metabisulfite has attracted a great deal of attention. It has the advantage of being a free-flowing granular material that is safe to handle and not particularly expensive. Used on fresh crops at the rate of 8–10 pounds per ton, varied but generally beneficial results have been obtained (31). The greatest benefits have been observed when treated silage has been compared to direct-cut silage of poor quality. In these cases significant decreases in dry matter loss and levels of butyric acid have been observed. This material has the same problem as most other additive materials; that is, the success of the additive is highly dependent upon the low quality of control silages (51).

Bacterial Cultures. Inoculation of silages with cultures of *Lactobacillus* has been repeatedly evaluated, particularly in Holland and England, as a means of improving silage quality. Results could best be described as variable; most of these investigations have been carried out in relatively small laboratory silos. McDonald et al. (31) found that the addition of inoculum containing eight strains of *Lactobacillus* did not improve the chemical characteristics of silage made from perennial ryegrass in pilot-sized silos. This forage initially contained about 16 percent total soluble carbohydrates. On the other hand, when the inoculum was applied to orchardgrass forage that contained only 4.3 percent total soluble carbohydrates, some advantage from the culture was noted. This British work, as well as some Dutch work (55), has indicated that the benefits of adding bacterial cultures are quite unpredictable, particularly when applied to farm-size silos and forages of reasonable carbohydrate content. The work of Langston et al. (26) suggests that the natural flora of most harvested forages is sufficiently variable to accommodate almost any type of fermentation. This work further suggests that the type of fermentation that does in fact develop depends mainly on the chemical composition of the crop and the storage conditions involved. Thus it seems unlikely that bacterial inoculations would have a very significant role in improving silage quality.

Acidification. Since it has long been recognized that the attainment of high-quality ensilage is largely dependent on the rapid development of a low pH, it is not surprising to find that the addition of strong acids to directly attain this end point has been considered by various workers since early in this century. The guiding principles and specific methods for direct acidification were quite thoroughly established by the Finnish nutritionist, A. I. Virtanen, starting about 1925. His method, which has become known as the AIV method, involves the addition of strong inorganic acids such as sulfuric and hydrochloric to forage at the time of ensiling to attain an immediate pH of less than 4. The concentrated acids are diluted with water to a final strength of 2 normal and sprinkled on the ensiled grass at the rate of 4–6 gallons per ton, depending on the forage species being ensiled. This method has been extensively tested in Europe and to a limited extent in the United States. There is general agreement that proper use of the AIV method will result in the elimination of respiration losses, the absence of butyric acid, and a very marked reduction in the degradation of protein.

Results with formic acid silage, reached by Waldo et al. (49), are encouraging. These workers reported a series of experiments in which direct-cut silage treated with 0.5 percent formic acid was compared to the same crop made into hay. The rate of growth for dairy heifers was consistently faster on the formic acid treated silage, and the efficiency of utilizing digestible energy was also consistently higher. In a subsequent report Waldo et al. (50) noted that formic acid treated silage had an advantage over untreated silage in terms of storage losses, intake levels, and efficiency of utilizing digestible energy. The cumulative effect of the several advantages was an increase of 50 percent in the pounds of gain per ton of forage initially stored. The value of this treatment for preserving forage for lactating cows has been variable (13).

Formic acid must be handled with care and involves some hazard to personnel. However, it presents no problem in terms of palatability or altered metabolism of animals. The mode of action of formic acid appears to be very similar to that noted with the AIV silage. This would explain the excellent preservation rates, low levels of butyric acid, and absence of protein hydrolysis. The method may prove to be very useful to producers who find wilting inconvenient. However, some of the problems of transporting and marketing formic acid must be solved before any large-scale acceptance can be expected in the United States.

FEEDING VALUE OF PRESERVED FORAGES

PUBLISHED INFORMATION on the comparative feeding value of preserved products is diverse and confusing. One can only conclude that quality variances within the same general preservation method are as great as between methods. Some of the apparent conflicts in the data are more easily explained if one keeps in mind the following facts: (1) the nutrition an animal obtains from forage is dependent on both the content of digestible nutrients and the amount consumed; (2) if the amount of feed is restricted, no expression of potentially different intake levels or palatability will be seen; (3) if forages are being fed ad libitum, the observed production levels may be fully as much a result of relative intake as relative digestible nutrient content.

The effects of preservation on the digestibility of forages have been studied for both ensiling and drying. There seems to be no sound basis for thinking that either process would increase digestibility. Therefore, a preserved product having digestibility equal to the fresh crop is considered optimum. Reduction of digestibility by ensiling has been reported in some studies (22), and in others digestibility equal to the fresh crop was observed (30). The loss of highly digestible solubles through seepage or deleterious heating may result in reduced digestibility. On the other hand, silage may be preserved without either of these events occurring. Therefore, the relative digestibility of fresh and ensiled crops is understandably variable. Similarly, digestibility of dried forage may or may not be depressed by slow drying or high temperatures.

When artificially dried, forage, hay prepared in several ways, and silage prepared in several ways have been compared on the basis of equalized intakes of dry matter or digestible nutrients; animal response has also been about equal (41, 49). This similarity of response supports the idea that major practical differences in feeding value result directly or indirectly from differentials in voluntary intake. Feeding comparisons made on the basis of unrestricted intakes, therefore, have more application to general farm practices.

The growth of young cattle has rather consistently been slower on control high-moisture silages compared to good quality hay (45, 48). The most prominent characteristic of these silages has been low intake levels, but the underlying cause of the low intake has not been fully identified. Gains equal to or higher than hay have been obtained by increasing silage dry matter content (17) and by formic acid treatments (49).

Dry matter consumption by lactating dairy cows tends to be lower on all silage rations than on all good quality hay and may result in lower milk production (9, 16). An improvement in consumption and production has been noted by prewilting the crop (17, 40). Several other workers (4, 6, 38, 43) have reported milk production as high or higher from all silage rations in spite of lower dry matter intakes. This type of relationship has suggested an increased utilization efficiency in the case of silages. Since lactating cows frequently utilize body stores as well as current energy to synthesize milk, estimates of true energetic efficiency by the usual feeding trial techniques are subject to large errors.

A discussion of the relative feeding value of various preserved forage products would be considerably simplified if improvements in preservation efficiency and chemical quality were invariably reflected in differential animal response. This, however, is certainly not the case, and almost any relative performance value for hay and silage may be found if the literature is searched diligently. It is best to keep in mind that efficient preservation and high chemical quality can to a large extent be justified on their own merits without regard to animal performance. Whether or not any superiority of animal performance per 100 units of preserved forage fed will be observed depends on the class of animals being fed, the level of expected production, the level of concentrate supplementation, the length of the feeding period, the crop species used, environmental temperature of the animals, and in some cases the number of times per day that the forage is offered. Therefore, efficient forage preservation may or may not be reflected in increased rates of animal performance.

A few factors should be mentioned because of their importance in determining the optimum method of preservation, even though they have very little to do with efficiency of method or animal response. When forage is being preserved for on-farm feeding under relatively humid conditions, a marked trend toward the use of silage may be noted. Complete mechanization of silage making and feeding is probably more responsible for the shift than any other single factor. Secondly, when forage is being prepared for sale and feeding in a location distant from the production area, the extra weight and spoilage possibilities practically preclude the use of silage for long distance hauls. In this situation, some form of field-dried hay is almost invariably used.

SUMMARY

THE PRESENT TREND toward preservation of a greater proportion of forage is likely to continue in areas adapted to intensive land use. Field-cured hay systems can be expected to continue on small farms and in areas of low rainfall. On large farms and in humid areas, greater emphasis on quality and mechanization suggests increased use of silage systems for on-farm fed forage, while a shift toward central dehydration and processing of forages would better serve the needs of distant drylot operations. Therefore the emergence of a single dominant system is not expected.

REFERENCES

1. Austenson, H. M. 1963. Influence of time of harvest on yield of dry matter and predicted digestibility of four forage grasses. *Agron. J.* 55:149–53.
2. Baxter, H. D., J. R. Owen, and D. R. Waldo. 1966. Effect of laceration of chopped forage on preservation and feeding value of alfalfa-orchard-grass silage. *J. Dairy Sci.* 49:1441–45.
3. Baxter, H. D., J. R. Owen, M. J. Montgomery, D. R. Waldo, and J. T. Miles. 1969. Pasturing vs. harvesting of a grass-legume mixture. Univ. Tenn. Agr. Exp. Sta. Bull. 454.
4. Brown, L. D., D. Hillman, C. A. Lassiter, and C. F. Huffman. 1963. Grass silage vs. hay for lactating dairy cows. *J. Dairy Sci.* 46:407–10.
5. Burns, J. C., C. H. Noller, and C. L. Rhykerd. 1964. Influence of methods of drying on the soluble carbohydrate content of alfalfa. *Agron. J.* 56:364–65.
6. Byers, J. H. 1965. Comparison of feeding value of alfalfa hay, silage, and low-moisture silage. *J. Dairy Sci.* 48:206–8.
7. Cole, G. L., W. E. McDaniel, and W. H. Mitchell. 1960. Hay drying costs and returns. Univ. Del. Agr. Exp. Sta. Bull. 334.
8. Colovos, N. F., N. K. Peterson, P. T. Blood, and H. A. Davis. 1965. The effect of rate of nitrogen fertilization, geographic location, and date of harvest on yield, acceptability, and nutritive value of timothy hay. Univ. N.H. Agr. Exp. Sta. Bull. 486.

9. Conrad, H. R., J. W. Hibbs, A. D. Pratt, and J. H. Vandersall. 1958. Milk production, feed intake, and digestibility following initiation of legume-grass silage feeding. *J. Anim. Sci.* 17:1197–98.

10. Darlington, J. M., and T. V. Hershberger. 1968. Effect of forage maturity on digestibility, intake, and nutritive value of alfalfa, timothy, and orchardgrass by equine. *J. Anim. Sci.* 27:1572–76.

11. Dawson, J. R., D. V. Kopland, and R. R. Graves. 1940. Yield, chemical composition, and feeding value for milk production of alfalfa hay cut at three stages of maturity. USDA Tech. Bull. 739.

12. De Man, J. G. 1952. Influence of crushing on the pH of grass silage. *Nature* 169:246–47.

13. Derbyshire, J. C., and C. H. Gordon. 1969. The utilization of formic acid silages by milk cows. USDA, ARS, DCRB 69-2 (mimeo).

14. Gordon, C. H. 1968. Loss of protein digestibility in stored forages, pp. 44–53. Proc. 1968 Maryland Nutrition Conf. for Feed Manufacturers.

15. Gordon, C. H., H. G. Wiseman, J. C. Derbyshire, W. C. Jacobson, and D. T. Black. 1959. Effect on silage of chopping and bruising the forage. *J. Dairy Sci.* 42:1394–95.

16. Gordon, C. H., J. C. Derbyshire, H. G. Wiseman, E. A. Kane, and C. G. Melin. 1961. Preservation and feeding value of alfalfa stored as hay, haylage, and direct-cut silage. *J. Dairy Sci.* 44:1299–1311.

17. Gordon, C. H., J. C. Derbyshire, W. C. Jacobson, and H. G. Wiseman. 1963. Feeding value of low-moisture alfalfa silage from conventional silos. *J. Dairy Sci.* 46:411–15.

18. Gordon, C. H., J. C. Derbyshire, W. C. Jacobson, and J. L. Humphrey. 1965. Effects of dry matter in low-moisture silage on preservation, acceptability, and feeding value for dairy cows. *J. Dairy Sci.* 48:1062–68.

19. Gordon, C. H., J. C. Derbyshire, C. W. Alexander, and D. E. McCloud. 1966. Effects of grazing pressure on the performance of dairy cattle and pastures, pp. 470–75. Proc. 10th Int. Grassl. Congr.

20. Gordon, C. H., R. D. Holdren, and J. C. Derbyshire. 1969. Field losses in harvesting wilted forage. *Agron. J.* 61:924–27.

21. Gouet, Ph., Nathalie Fatianoff, S. Z. Zelter, Michelle Durand, and R. Chevalier. 1965. Influence de l'elévation du taux de matière sèche sur l'evolution biochimique et bactériologique d'une luzerne conservée par ensilage. *Ann. Biol. Anim. Bioch. Biophys.* 5:79–100.

22. Harris, C. E., and W. F. Raymond. 1963. The effect of ensiling on crop digestibility. *J. Brit. Grassl. Soc.* 18:204–12.

23. Hart, R. H., and G. W. Burton. 1967. Curing Coastal bermudagrass hay: Effects of weather, yield, and quality of fresh herbage on drying rate, yield, and quality of cured hay. *Agron. J.* 59:367–71.

24. Hoglund, C. R. 1967. Changes in forage production and handling on southern Michigan dairy farms. Mich. State Univ. Agr. Econ. Rept. 78.

25. Lancaster, R. J., and Mary McNaughton. 1961. Effects of initial consolidation on silage. *N.Z. J. Agr. Res.* 4:504–15.

26. Langston, C. W., H. Irvin, C. H. Gordon, Cecelia Bouma, H. G. Wiseman, C. G. Melin, L. A. Moore, and J. R. McCalmont. 1958. Microbiology and chemistry of grass silage. USDA Tech. Bull. 1187.

27. Larsen, H. J. 1959. Methods of forage utilization in the Midwest. *J. Dairy Sci.* 42:574–78.

28. Linnerud, A. C., J. D. Donker, J. Strait, and A. M. Flikke. 1961. A report of hay treatment studies and feeding trials. *Minn. Farm and Home Sci.* Vol. 43.

29. Longhouse, A. D. 1960. Hay conditioners in the northeastern United States. W. Va. Univ. Agr. Exp. Sta. Bull. 449.

30. McDonald, P., A. C. Stirling, A. R. Henderson, W. A. Deware, G. H. Stark, W. G. Davie, H. T. Macpherson, A. M. Reid, and J. Slater. 1960. Studies on ensilage. The Edinburgh School of Agr. Tech. Bull. 24.

31. McDonald, P., A. C. Stirling, A. R. Henderson, and R. Whittenbury. 1964. Fermentation studies on inoculated herbages. *J. Sci. Food & Agr.* 15:429–36.

32. Miller, L. G., D. C. Clanton, L. F. Nelson, and O. E. Hoehne. 1967. Nutritive value of hay baled at various moisture contents. *J. Anim. Sci.* 26: 1369–73.

33. Minson, D. J., C. E. Harris, W. F. Raymond, and R. Milford. 1964. The digestibility and voluntary intake of S22 and H1 ryegrass, S170 tall fescue, S48 timothy, S215 meadow fescue, and germinal cocksfoot. *J. Brit. Grassl. Soc.* 19:298–305.

34. National Academy of Sciences. 1966. Nutrient requirements of dairy cattle. Nat. Res. Council. Publ. 1349, Washington, D.C.

35. Nelson, L. F. 1966. Spontaneous heating and nutrient retention of baled alfalfa hay during storage, pp. 509–12. Trans. ASAE.

36. Nilsson, Gerda. 1959. Silage studies. 9. Biochemical changes in microbe-free silage. *Archiv für Mikrobiologie* 34:30–35.

37. Poulton, B. R., M. J. Anderson, and T. N. Melin. 1961. Nutritive value of hay depends on cutting date. *Maine Farm Research* (Jan.).

38. Pratt, A. D., and H. R. Conrad. 1957. Proportions of hay and silage in the dairy ration. *J. Dairy Sci.* 40:620.

39. Reid, J. T., W. K. Kennedy, K. L. Turk, S. T. Slack, G. W. Trimberger, and R. P. Murphy. 1959. Effect of growth stage upon the nutritive value of forages. *J. Dairy Sci.* 42:567–71.

40. Roffler, R. E., R. P. Niedermeier, and B. R. Baumgardt. 1967. Evaluation of alfalfa-brome forage stored as wilted silage, low-moisture silage, and hay. *J. Dairy Sci.* 50:1805–13.

41. Shepherd, J. B., H. G. Wiseman, R. E. Ely, C. G. Melin, W. J. Sweetman, C. H. Gordon, L. G. Schoenleber, R. E. Wagner, L. E. Campbell, G. D. Roane, and W. H. Hosterman. 1954. Experiments in harvesting and preserving alfalfa for dairy cattle feed. USDA Tech. Bull. 1079.

42. Shepherd, J. B., R. E. Ely, C. H. Gordon, and C. G. Melin. 1956. Permanent pasture compared with a 5-year crop-and-pasture rotation for dairy cattle feed. USDA Tech. Bull. 1144.

43. Slack, S. T., W. K. Kennedy, K. L. Turk, J. T. Reid, and G. W. Trimberger. 1960. Effect of curing methods and stage of maturity upon feeding value of roughages. 2. Different levels of grain. Cornell Univ. Agr. Exp. Sta. Bull. 957.

44. Stallcup, O. T., G. V. Davis, and D. A. Ward. 1964. Factors influencing the nutritive value of forages utilized by cattle. Univ. Ark. Agr. Exp. Sta. Bull. 684.

45. Thomas, J. W., L. A. Moore, and J. F. Sykes. 1961. Further comparisons of alfalfa hay and alfalfa silage for growing dairy heifers. *J. Dairy Sci.* 44:862–73.

46. Van Riber, G. E., and Dale Smith. 1959. Changes in the chemical composition of the herbage of alfalfa, medium red clover, ladino clover, and bromegrass with advance in maturity. Univ. Wis. Agr. Exp. Sta. Res. Rept. 4 (Sept.).

47. Van Soest, P. J. 1965. Use of detergents in analysis of fibrous feeds. 3. Study of effects of heating and drying on yield of fiber and lignin in forages. *J. AOAC* 48:785–90.

48. Waldo, D. R., R. W. Miller, L. W. Smith, M. Okamoto, and L. A. Moore. 1966. The effect of direct-cut silage, compared to hay, on intake, digestibility, nitrogen utilization, heifer growth, and rumen retention, pp. 570–74. Proc. 10th Int. Grassl. Congr.

49. Waldo, D. R., L. W. Smith, R. W. Miller, and L. A. Moore. 1969. Growth, intake and digestibility from formic acid silage versus hay. *J. Dairy Sci.* 52:1609–16.

50. Waldo, D. R., J. E. Keys, Jr., L. W. Smith, and C. H. Gordon. 1971. Effect of formic acid on recovery, intake, digestibility and growth from unwilted silage. *J. Dairy Sci.* 54 (1): 77–84.

51. Watson, S. J., and M. J. Nash. 1960. The Conservation of Grass and Forage Crops, 2nd ed. Oliver and Boyd, Edinburgh and London.

52. Whittenbury, R., P. McDonald, and D. G. Bryan-Jones. 1967. A short review of some biochemical and microbiological aspects of ensilage. *J. Sci. Food Agr.* 18:441–44.

53. Wieringa, G. W. 1958. The effect of wilting on butyric acid fermentation in silage. *Neth. J. Agr. Sci.* 6:204–10.

54. Wieringa, G. W., S. Schukking, D. Kappelle, and S. J. De Haan. 1961. The influence of heating on silage fermentation and quality. *Neth. J. Agr. Sci.* 9:210–16.

55. Wieringa, G. W., and A. G. Hengeveld. 1963. Inoculation with lactic-acid bacteria, and adding sugar when ensiling. *Landbouwvoorlichting* 20:587–92 (Dutch).

56. Wysong, J. W. 1961. Costs, returns, and profitability of artificial hay drying on northeastern dairy farms. Univ. Md. Agr. Exp. Sta. Bull. 471.

57. Zimmer, Ernst, and C. H. Gordon. 1964. Effects of wilting, grinding, and aerating on losses and quality in alfalfa silage. *J. Dairy Sci.* 47:652–53.

7

ROLE OF RUMINANT LIVESTOCK IN MEETING THE WORLD PROTEIN DEFICIT

L . A . M O O R E , P . A . P U T N A M , *and*
N . D . B A Y L E Y

WORLD LEADERS recognize the potential tragedy of mass starvation if increases in food production do not keep pace with increases in population during the next 50 years. From a nutritional viewpoint, protein is expected to be the foodstuff to become limited first. In 35 countries of the three major underdeveloped regions of the world, the average per capita consumption of animal, pulse, or total protein is below minimum recommended levels according to the World Food Budget. The population of these countries represents 79 percent of the population of the regions and 56 percent of the world population (2).

Studies on the contribution of U.S. agriculture in supplying world protein needs have proposed emphasis on increased production of properly supplemented cereals and high-protein oilseeds for direct use as food for humans (2). Development of other protein sources such as algae, leaves, yeast, fungi, and bacteria has been suggested (1, 27). Meeting these protein deficits with animal products has been considered impractical (20).

Although the emphasis on cereal and oilseed proteins has some basis, relegating animal agriculture to a passive contribution to world food deficits indicates a failure to appreciate the full impact of feed inputs into livestock production. We contend that generally accepted concepts regarding the efficiency of livestock production in terms of use of available resources are erroneous. We contend that because livestock use forages and other feeds inedible to humans, the use of limited amounts of cereals as livestock feeds can enhance the efficiency of producing proteins for humans *in terms of total food*

Used with permission from *Agr. Sci. Rev.*, Vol. 5, No. 2, Second Quarter, 1967.

resource utilization. Furthermore, promising research leads, if exploited, can markedly increase the efficiency with which animal proteins can be produced. We also contend that considering the world food deficits solely in terms of amounts of protein or calories may result in answers which will make only the less desired diets available to the "have-nots" and may aggravate the serious sociological problems of the world rather than reduce them.

LAND AND FOOD RESOURCE UTILIZATION

WHY HAVE INCREASES in cereal and oilseed protein production been emphasized? The partial answer can be found in the statistical problems of making world food studies. Indices of crop production can be based on reasonably reliable estimates of yield assembled by acceptably uniform methods from country to country. Indices on livestock and livestock products are not practical because reliable estimates of yield do not exist in many countries. Also, there are real difficulties in taking into account the feed-grain imports and the feed-grain transfers from the crop to the livestock economy (24). The use of crop yields, particularly grains considered aggregately, also greatly simplifies the analysis (2). Therefore, because of the relative ease of collecting and compiling crop production data, the major studies on present and potential food production are based on grain yields and ignore livestock production.

The second reason for emphasizing grains in these studies is based on the importance of grains as a source of man's food. For example Brown (2) states, "Grains account for 71 percent of the world's harvested crop area; they provide 53 percent of man's supply of food energy when consumed directly and a large part of the remainder when consumed indirectly in the form of livestock products." These statements are factual, and the dominant role of grains in producing human food directly or indirectly is indisputable.

However, the next basis for emphasizing cereal and oilseed crops in world food protein recommendations is open to question. The efficiency with which livestock convert protein from cereals and oilseed into meat and milk proteins is often quoted as about 10 percent and 25 percent respectively (27). The implication from this low efficiency of conversion is that the feeding of cereals and oilseeds to livestock to produce proteins for human food is a wasteful practice and livestock compete directly with humans for proteins needed. This implication, however, is not based on all the facts in livestock feeding. It would be true only if livestock were exclusively grain-consuming animals. This assumption is not valid.

Farm animals, particularly the ruminants, have the ability to convert feeds other than cereals and oilseeds into protein for human consumption. These other feeds consist of forages, by-products from the harvesting or processing of food crops, by-products from the processing of animal products, and nonprotein sources of nitrogen.

Forages are food resources not generally consumed directly by humans. However, forages make up a large proportion of the rations of farm animals. About 70 percent of the protein of the average U.S. dairy cow is obtained from forages. An average beef steer gets 60 percent of its protein from forages; sheep from 80 to 90 percent. Swine and poultry, of course, use much less forage. In other countries of the world, the percentage of forage in the ration is considerably higher.

Forages, furthermore, are often grown on land which is submarginal or entirely unsuitable for producing other crops. In the United States about 191 million hectares (472 million acres) of permanent grassland pasture and range are either privately owned or owned by public agencies other than the federal government (28). The productivity of this land is only one-fifth to one-sixth that of most cropland. However, the immense acreage makes this land an important food source when converted into animal proteins by livestock (3).

In addition, there are 162 million hectares (400 million acres) of federally owned grassland range and federally owned (and otherwise) forests and woodland pasture and range. The productivity of this land is only 3–4 percent of that of average cropland. However, the importance of it as a feed source is evident from the fact that one-eighth of the livestock feed in the mountain and Pacific states comes from federal range alone (3). About 50 percent of the sheep and 20 percent of the beef cattle in the United States are in these states.

Trends in the United States toward greater grain consumption and less forage consumption by livestock are sometimes interpreted as irreversible results of new technology. These trends have been used as evidence to support the belief that livestock increasingly compete with humans for food. Such interpretations are not correct. The greater use of grains by livestock is a result of new technologies that have been making grains increasingly abundant and inexpensive as livestock feeds. The economic advantage of using grains is the reason for the trends. If the demand for grains as human food becomes sufficient to raise their price beyond economic levels for livestock feeds, these trends will reverse themselves rapidly. The importance of forages in livestock feeding can be expected to increase rather than decrease, particularly if the overall human food supply becomes as critical as many are predicting.

All classes of livestock use feeds which are by-products of the manufacture of flour, starch, glucose, and other human food items. Some of the more important by-product feeds are wheat mill feeds, corn gluten feed and meal, distillers dried grains, and rice mill feeds. From 1950 to 1959 the annual animal consumption of these feeds ranged from 10 to 11.8 million metric tons (11.9–13.0 million tons) (9). Classed as low and medium protein feeds, most have higher protein content than the grains from which they are derived. These feeds are not suitable for direct human consumption but can be converted into nutritious and palatable milk and meat proteins by feeding to livestock. Even though these feeds are low in protein, they provide energy needed in the production of animal proteins. One of the more important of these is dried sugar beet pulp made from extracted sugar beets. Approximately 169,000 metric tons (186,000 tons) are annually consumed by livestock in the United States. Wet beet pulp is also fed in areas adjacent to processing plants. Another by-product from the same source, beet syrup, is also fed in large amounts.

In the meat packing industry, inedible portions of carcasses including meat scraps, intestines, and blood are processed by high-temperature rendering into livestock feeds. About 2.1 million metric tons (2.3 million tons) of such animal by-products were fed to livestock in 1963.

Urea, a nonprotein nitrogen source which humans cannot utilize in their diet, can be converted to animal proteins by ruminants. It is being used extensively in dairy and beef cattle feeding and to a lesser extent in lamb rations (10). Approximately 127,000 metric tons (140,000 tons) of urea are being fed to livestock annually. If soybean meal had been fed in place of urea, the requirements for soybean meal for livestock feed would have been increased by 813,000 metric tons (896,000 tons).

The toxicity and palatability of urea is a factor in its proper utilization. However, the main restraint on increased use of urea is the availability and low price of large amounts of feed grains and

TABLE 7.1. *Protein inputs-outputs from cereals and oilseeds in a dairy cow operation*

Source of input	Crude protein			
	Input		Output in milk	
	(kg)	*(lb)*	*(kg)*	*(lb)*
1439 kg grain and oilseed concentrates per year	178	(391.6)	171	(376.2)

Note: Data for cows producing 5295 kg (11,649 lb) of milk containing 3.69 percent butterfat.

oilseeds. Use of urea can be expected to increase, particularly if the demand for food proteins from cereals and oilseeds markedly increases the price of these crops.

Therefore, before concluding that livestock compete with humans for the proteins from cereals and oilseeds, one must consider the proteins that livestock produce from other sources.

ANIMAL INPUT AND OUTPUT OF PROTEIN

IN A WISCONSIN survey (5) of 1500 dairy cows in 46 herds, 96 percent of the protein in their cereal and oilseed rations was returned as milk protein (Table 7.1). Total protein input has been corrected for crop by-products inedible to humans.

If urea had been used at recommended levels in the ration, the cows would have produced 171 kg (377 lb) of milk protein from only 124 kg (273 lb) of plant proteins in the form of cereals and oilseeds. Thus protein production would have been increased nearly 40 percent by combining cereals and oilseeds with forages and urea to produce milk.

Table 7.2 shows the protein inputs from grains and oilseeds and the outputs from a beef cow-calf operation (15). An 80 percent calf crop per year and a weaning weight of 181 kg (399 lb) for calves were used as bases for these estimates. Although below expectations for good management practices, these bases are probably somewhat above the actual average for the United States. These data indicate that feeding cereals and oilseeds to beef cattle results in inefficient use of plant proteins. However, the management and performance levels in Table 5.2 were deliberately chosen below recommended practices.

Table 7.3 shows what is possible in beef cattle production if recommended range and pasture management practices and improved

TABLE 7.2. *Protein inputs-outputs from cereals and oilseeds in a beef cow-calf operation*

Feeding period	Crude protein			
	Inputs		Outputs	
	(kg)	(lb)	(kg)	(lb)
Birth to 181 kg (398.2 lb)	0		0	
181 kg to weaning of 1st calf	27	(59.4)	14	(30.8)
To weaning of 2nd calf	14	(30.8)	14	(30.8)
To weaning of 3rd calf	14	(30.8)	14	(30.8)
To weaning of 4th calf	14	(30.8)	14	(30.8)
Barren	14	(30.8)	. . .	. . .
Slaughtered	. . .	. . .	34	(74.8)
Calves fed out to 454 kg (998.8 lb)	392	(862.4)	67	(147.4)
Total	475	(1045.0)	157	(345.4)

Note: Based on 80 percent annual calf crop and 181 kg (398.1 lb) weaning weight.

breeding are combined for greater efficiency (8, 9, 10, 15, 17). The beef cow produced seven calves before being slaughtered, and the weaning weights were 277 kg (610.8 lb). Such performance levels are attainable by progressive beef producers. Also, urea was substituted for grains and oilseeds in accordance with recommendations.

Data in Table 7.3 indicate that with good breeding and management and the use of urea, nearly as many kilograms (pounds) of beef proteins can be obtained as the kilograms (pounds) of grain and oilseed proteins provided to the cattle. In addition some cattle are grass fattened without the usual heavy feeding of concentrates. The protein yield from these animals would greatly exceed the pro-

TABLE 7.3. *Protein inputs-outputs from cereals and oilseeds in a beef cow-calf operation*

Feeding period	Crude protein			
	Inputs		Outputs	
	(kg)	(lb)	(kg)	(lb)
Birth to 227 kg* (499.4 lb)	0		0	(35.2)
227 kg to weaning of 1st calf	0		16	(35.2)
To weaning of 2nd calf	0		16	(35.2)
To weaning of 3rd calf	0		16	(35.2)
To weaning of 4th calf	0		16	(35.2)
To weaning of 5th calf	0		16	(35.2)
To weaning of 6th calf	0		16	(35.2)
To weaning of 7th calf	0		16	(35.2)
Slaughtered	0		34	(74.8)
Calves fed to 454 kg† (998.8 lb)	249	(547.8)	98	(215.6)
Total	249	(547.8)	244	(536.8)

Note: Based on 7 calves per cow and 227 kg (498.4 lb) weaning weight.
* Winter protein supplement as urea-molasses.
† Corn and cob meal with urea as supplementary nitrogen.

tein provided them from grains and oilseeds. Furthermore, no credit is given in these tables for corn by-products such as distillery grains or corn gluten, which are not used by humans but could be incorporated into rations for beef cattle.

These examples illustrate the importance of dairy and beef cattle in the efficient production of proteins for human use. This, however, is only part of the story. Promising research leads indicate that the efficiency with which cattle and sheep utilize nonprotein nitrogen (NPN) combined with low-quality forages can be greatly increased. Some reports have shown that NPN can serve as the sole source of nitrogen in the rations of ruminants. If such a development could be placed into practical use, it would mean that no exogenous source of protein would be required for the production of animal products by ruminants.

NONPROTEIN NITROGEN

IN 1949 it was shown that the essential amino acids could be synthesized in the rumen from urea (13). In experiments conducted in Finland (25), dairy cows fed a ration of urea, potato starch, cellulose, and sucrose have produced in one year up to 4325 kg (9515 lb) of milk and 164 kg (360.8 lb) of protein without any feed source of protein. In experiments conducted at Beltsville (18), an Angus female weighing 132 kg (290.4 lb) was put on a purified diet containing urea and no feed source of protein. She gained at the rate of 0.45 kg (.9 lb) per day to a weight of 422 kg (928.4 lb) and produced a calf. In experiments with lambs, an average daily gain of 0.10 kg (.22 lb) was obtained with urea as a sole source of nitrogen, while isolated soy protein produced 0.12 kg (.264 lb) gain (19).

Although the literature clearly indicates that ruminants can produce meat and milk without any exogenous source of protein, further research is needed to increase the level of production of meat and milk from nonprotein sources. One of the difficulties is related to the timing of the release of ammonia nitrogen in the rumen from the NPN source and the release of energy from cellulose when forage is the primary energy source. The ammonia N from urea is released rapidly after it is ingested, but the energy from forages is released slowly. In order to promote synthesis of amino acids, the release time of each must be synchronized in the rumen.

That slowing the availability of ammonia N in the rumen will aid in its utilization is demonstrated by the fact that feeding urea six times compared to two times daily will improve gain in dairy heifers (4). Possible ways to slow the rate of release of ammonia N in the rumen would be to (1) pelletize the urea, or (2) coat pelleted urea with a slowly dissolving material. Another possibility would be to develop other ammonia N compounds or compounds related to urea which would release nitrogen more slowly in the rumen. The possible use of clathrates of urea or other ammonia N bearing compounds may prove useful. The effect of pelleting the clathrates should also be investigated.

HIGH-FIBER FEEDS

IN ADDITION to improving the utilization of urea, the treatment of low-quality forages themselves to increase their available energy would further improve the efficiency of livestock production. On the basis of projected increases in crop production in the United States, straw products alone could support an appreciable part of an increased animal population. For example, in the production of corn, wheat, etc., to feed the human population, about one-half the dry matter is in the form of straw and is not utilized as food or as animal feed. On a total digestible nutrient basis it is estimated that by 1980 (7) this source of energy from grain production in the United States would be sufficient to maintain a herd of 49 million dairy cows and to produce 4500 kg (9900 lb) of milk per cow per year. The present national dairy herd has about 15 million cows.

In their present form, poor quality forages such as straw have been of limited value to ruminants because of low digestibility. Furthermore, the lower the digestibility of forage because of high fiber content, the slower the rate of digestion. Likewise, the slower the rate of digestion in the rumen, the slower the rate of passage from the rumen and the lower the intake. Methods should be developed to improve the intake and utilization of poor quality forages. Guidance in the chemical treatment of poor quality forages such as straw to make the cellulose more available can be obtained from the vast literature of the wood pulping industry. Finnish workers (21), using rams, have examined the effect of wood pulping treatments on the digestibility of spruce or birchwood. Digestibility of the crude carbohydrate fraction varied between 27.5 and 89.8 percent depending upon the treatment. It is estimated that the digestibility of wood before the treatment would be only about 9 or 10 percent. The technical cellulose pulp produced has a feeding value of 0.9 feed units or is equivalent to 90 percent that of barley.

It has been shown that the treatment of basswood by high velocity electrons alters its structure in such a way that some of the insoluble carbohydrate components become available to rumen bacteria (12). Extracellular enzymes in certain fungi are known to break down lignin (14), and might be useful in the delignification of forages to make the cellulose more readily available to the animal. The lignin content of wood is usually 20 percent or higher (21), whereas the lignin content of straw is only 7–8 percent. This lower lignin content and the consequent need for much less drastic chemical treatment could make the delignification of straw more economically feasible than wood. In experiments conducted in Norway (11) during World War II, straw was treated with 1.2–1.5 percent sodium hydroxide. The digestibility of the organic matter was increased from 42.4 to 65.7 percent. Along this line, data from Beltsville (26) indicate the possibility of increasing the digestibility of forages by as much as 10 digestibility units by certain ensiling procedures. The digestion coefficient (22) for wheat straw was increased from 22.8 to 52.2, as determined by an artificial rumen digestion technique, through treatment of straw with chlorine dioxide.

In chemically treating poor quality forages to increase their digestibility, the soluble constituents probably would be recovered in the form of molasses. These constituents would be practically 100 percent digestible and would form a readily available source of energy. They could also probably be used to improve the palatability of the treated forages.

Using treated wood as a feed for animals is presently limited by economic barriers. If these barriers should change, however, wood could become a means for increasing the efficiency of livestock production. Furthermore, the possibility exists for recovering the soluble nutrients from the wood pulping industry as a molasses feed and thus preventing stream pollution. The feeding value of wood material in the form of molasses as a product of cellulose manufacture has already been examined in feeding trials with steers (6).

MANAGEMENT NEEDS

IN ADDITION to research on the use of nonprotein nitrogen and the digestibility of low-quality forages, the potential of research on improving forage production should not be overlooked. This is particularly pertinent to developing countries where present practices in utilizing forages for protein production are extremely inadequate. In some countries, cattle are on deficit rations during the dry months and have adequate nutrition only during a few months of heavy rainfall.

We do not imply that future rations of ruminants will consist only of urea and chemically treated poor quality forages. Basic feeds are likely to continue to be cereals, cereal by-products, and good quality forages. But as competition for food for man increases, economics will dictate the extent to which urea and poor quality forages will be used.

SOCIOLOGICAL RELATIONSHIPS

IF THE CONTRIBUTION animal products can make to solving world protein problems is arbitrarily held at a minimum and thus not fully realized, the solution to the major problem—the sociological relationship between the "haves" and "have-nots"—could be greatly aggravated. Admittedly, plant proteins lower in quality than animal proteins can be improved by combining various plant sources or by supplementing them with synthesized amino acids. Admittedly, proteins from some of these plant sources can be provided at lower cost than animal proteins. But the fact remains that most people prefer to eat animal products except where consumption is forbidden by law or social custom.

The FAO food report (23) shows that countries with the higher per capita incomes are generally the largest consumers of animal products. The availability of these products is one of the symbols of affluence to which people in underdeveloped nations aspire. Although their minimum nutritional requirements may be met by less than desired protein sources, their sociological needs will not be met by having to exist on second-choice foods. As incomes rise, the demand for more and more animal products will obviously also rise. Japan has already demonstrated that countries are ready and willing to pay the price for animal proteins even when they have to import such grains as are needed to supplement existing feedstuffs (16).

SUMMARY

AS THE STANDARD of living of people in underdeveloped countries increases, it is difficult to imagine that they are not going to demand more than the minimum quality dietary fare. Planning by the United

States to meet the needs of developing countries must provide for increased production of animal products. The potential for this increase exists in the great areas of the world's grasslands and in improved management of these and other lands too rough, too dry, or too infertile for cropping. The challenge to animal agriculture is to assert its proper role in the production of world proteins by increasing the understanding of world leaders regarding the real efficiency of livestock production and to conduct the research needed in order to further improve that efficiency.

REFERENCES

1. Altschul, A. M. 1965. Proteins, Their Chemistry and Politics. Basic Books, New York.
2. Brown, L. R. 1963. Man, Land, and Food. USDA Foreign Agr. Econ. Rept. 11.
3. Byerly, T. C. 1966. The role of livestock in food production. *J. Anim. Sci.* 25:552.
4. Campbell, J. R., W. M. Howe, F. A. Martz, and C. P. Merilan. 1963. Effects of frequency of feeding on urea utilization and growth characteristics in dairy heifers. *J. Dairy Sci.* 46:131.
5. Corley, E. L., G. R. Barr, D. A. Wieckert, E. E. Heizer, and C. E. Kraemer. 1964. Environmental influences on production in 46 dairy herds, pp. 1–64. Univ. Wis. Res. Bull. 253.
6. Cullison, A. E. 1966. Noncellulosic wood carbohydrates in steer rations, pp. 2–26. Proc. Ga. Nutr. Conf.
7. Daly, R. F., and A. C. Egbert. 1966. A look ahead for food and agriculture. *Agr. Econ. Res.* 18:1.
8. Duncan, H. R. 1958. Producing beef on grass from yearling and two-year-old steers with and without supplemental feeds. Univ. Tenn. Agr. Exp. Sta. Bull. 283.
9. Hodges, E. F. 1964. Consumption of feed by livestock, 1940–1959. USDA Res. Rept. 79.
10. ———. 1965. Feed use of urea in the United States. USDA Admin. Rept.
11. Hvidsten, H., and T. Homb. 1947. A survey of cellulose and Beckman-treated straw as feed, pp. 1–14. Roy. Agr. Coll. Norway Rept. 62.
12. Lawton, E. J., W. D. Bellamy, R. E. Hungate, M. P. Bryant, and E. Hall. 1951. Some effects of high-velocity electrons on wood. *Science* 113:380.
13. Loosli, J. K., H. H. Williams, W. E. Thomas, F. H. Ferris, and L. A. Maynard. 1949. Synthesis of amino acids in the rumen. *Science* 110:144.
14. Morira, E., and K. Henderson. 1963. Fungal metabolism of certain aromatic compounds related to lignin. *Pure Appl. Chem.* 7:589.
15. Nelson, A. G. 1945. Relation of feed consumed to food products produced by fattening cattle, pp. 1–36. USDA Tech. Bull. 900.
16. Novotny, D. 1960. Japan's livestock boom. USDA, ERS. Foreign Agr.

17. Oltjen, R. R., R. E. Davis, and R. L. Hiner. 1965. Factors affecting performance and carcass characteristics of cattle fed all-concentrate rations. *J. Anim. Sci.* 24:192.

18. Oltjen, R. R., J. Bond, P. A. Putnam, and R. E. Davis. 1965. First calf born on a synthetic diet, p. 8. USDA Agr. Res. Rept. 13.

19. Oltjen, R. R., R. J. Sirny, and A. D. Tillman. 1962. Purified diet studies with sheep. *J. Anim. Sci.* 21:277.

20. Paarlberg, D. 1966. The potential impact on animal agriculture of a changing U.S. policy as a result of the world's food population ratio. Nat. Inst. Anim. Agr., Purdue Univ., Lafayette.

21. Saarinen, P., W. Jensen, and J. Alhojarvi. 1959. On the digestibility of high yield chemical pulp and its evaluation. *Suomen Maataloustieteellisen Seuran Julkaisuja* 94:41.

22. Sullivan, J. T., and T. V. Hershberger. 1959. Effect of chlorine dioxide on lignin content and cellulose digestibility. *Science* 130:1252.

23. United Nations. 1964. FAO Production Yearbook 1963, Vol. 17.

24. USDA. 1965. Changes in agriculture in 26 developing nations. USDA Foreign Agr. Econ. Rept. 27.

25. Virtanen, A. I. 1966. Milk production of cows on protein-free feed. *Science* 153:1603.

26. Waldo, D. R., L. W. Smith, R. W. Miller, and L. A. Moore. 1966. Formic acid silage versus hay for growth. *J. Anim. Sci.* 25:916.

27. Weiss, M. G., and R. M. Leverton. 1964. World sources of protein. *In* Farmer's World, p. 44. USDA Yearbook Agr.

28. Wooten, H. H., K. Gertel, and W. C. Pendleton. 1962. Major uses of land and water in the United States, pp. 1–54. USDA Agr. Econ. Rept. 13.

8

FORAGES AND GRASSLANDS IN THE NORTHEAST

J O H N   B .   W A S H K O

THE NORTHEAST extends from the state of West Virginia in the south to Maine in the north (Fig. 8.1). The twelve states of this region occupy approximately 15 percent of the land area of the continental United States. The Northeast is considered primarily an industrial area; its agricultural importance is frequently overlooked. The region's agricultural history dates back to Colonial times when the colonists grew food for themselves and forages (pasturage during the growing season and hay for the long New England winters) for the livestock they brought with them.

The first pastures in the English colonies were natural clearings in the lowlands along the banks of streams and woods where the underbrush was burned by the Indians to aid hunting. The colonists found two native forage plants, wild rye and broomstraw (5). Early commentators wrote enthusiastically about these grasses, which grew tall and thick. Cattle ate them fully during the summer, but it was impossible to make enough hay to carry the livestock through the winter. Early settlers tried to make additional hay from the coarse reeds and sedges of the freshwater and saltwater marshes during the first half of the seventeenth century.

Although the colonists brought grass and clover seed with them from England, none of the early introduced species became widespread in the region. It was not until about 1700 that John Herd found a grass near Portsmouth, New Hampshire, that was to become an important forage species in the colonies. This grass, timothy, spread southward through the colonies, with the help of Timothy Hansen after whom it was named, and became the leading cultivated forage grass because of its wide adaptation. Forages continued to

98

FIG. 8.1.   THE NORTHEAST REGION (OUTLINED IN BLACK).

grow in importance within the region so that by the time the first agricultural census was taken in 1839, the main hay-producing area in the United States was in the northeastern states.

UTILIZATION OF LAND RESOURCES IN THE REGION

TOTAL LAND AREA in the region comprises 127.6 million acres (2). Forest and woodland vegetation predominates in the Northeast, presently occupying 62.9 million acres. Farms occupy another 41.2 million acres of the region. Of this farm acreage, only 14.3 million acres are classed as harvested cropland because economic, climatic, soil, and topographic factors are such that large acreages of land in the Northeast are either poorly suited or totally impractical for cultivated crops. However, much of this land area is adapted for forage and pasture production. In fact, 42.2 percent of the cropland harvested, approximately 7.5 million acres, is used for the production of grass silage and hay (Table 8.1) (4). An additional 12.8 million acres of land within the region are used for pasture. Combining the number of cropland acres harvested as forage crops with the pasture acreage indicates that grasslands occupy 20.3 million acres or 49.3 percent of northeastern farmland. Another 1.1 million acres of cropland are

TABLE 8.1. *Acreage devoted to field and forage crops in the Northeast*

State	Corn for grain	Corn for green forage	All other grains*	Potatoes	Silage Corn	Silage Grass	All hay	All pastures	Total in forage production†
					(000 acres)				
Maine	...	...	40	131	10	9	355	370	744
New Hampshire	...	...	...	2	10	4	140	219	373
Vermont	1	1	11	2	41	24	560	996	1,621
Massachusetts	...	...	1	6	23	12	147	239	422
Rhode Island	...	...	1	5	4	2	14	25	45
Connecticut	1	1	2	6	33	9	126	222	391
New York	178	16	735	74	427	134	2,750	4,109	7,436
New Jersey	57	6	126	17	49	13	148	191	407
Pennsylvania	777	15	1,117	35	333	115	1,972	2,778	5,213
Delaware	151	1	191	8	10	2	29	53	95
Maryland	394	10	455	3	105	26	334	678	1,153
West Virginia	52	3	45	3	31	10	543	2,875	3,462
Total	1,611	53	2,724	292	1,076	360	7,118	12,755	21,362

Source: USDC, 1966 (4).
Note: Acreages are to the nearest 1000 acres.
* Includes soybeans grown for grain.
† Includes acreages in corn green chop, corn and grass silage, hay and pasture.

planted to corn each year for silage and green chop. Therefore, approximately 21.4 million acres of the 41.2 million acres, or 51.9 percent, of the area's farmland are committed to forage and pasture production for livestock. Grain production is limited to approximately 4.3 million acres, or only 10.4 percent of the region's farmland. The growing of grain is confined primarily to areas where it is best adapted within the region to supplement the feed produced by grasslands.

Except in the southern part of the region where climatic conditions favor the production of small grains and corn, insufficient grain is grown to meet the demands of grain-consuming livestock. As indicated in Table 8.1, grain production is limited in the upper New England states because of severe winters and cool short summers. The Northeast, therefore, is considered a grain-deficient region which depends heavily on the Midwest to supply its grain needs and concentrates.

Potatoes are grown as a cash crop within each state of the Northeast on a total of 292,000 acres. When the acreages of northeastern field crops are compared, it becomes apparent why the area has been agriculturally designated as part of the hay and pasture region of the United States.

HARVESTED FORAGE

NORTHEASTERN FARMERS harvest approximately 7.1 million acres of hay each year (Table 8.2) (4). Red clover in grass mixtures is the leading hay legume despite its short life span of two years within the region. Apparently this is due to its wide adaptation within the area and its place in short rotations. Timothy is another well-adapted species within the region, and it is widely teamed with red clover for hay. As indicated in Table 8.2, the 3.8 million acres of red clover in grass mixtures exceeds that of alfalfa-grass mixtures and other forage species used for hay within all 12 states. The two largest agricultural states within the region, New York and Pennsylvania, each grow over 1 million acres of red clover–grass for hay annually. In 1964 the 12 northeastern states produced 5.8 million tons of red clover–grass hay.

Alfalfa alone and in grass mixtures was grown on approximately 2.5 million acres in the region in 1964. Alfalfa is more exacting in its soil requirements, particularly drainage, and must be confined to the better soils in the region. The tall-growing perennial grasses with

TABLE 8.2. *Acreage in types of hay and pasture in the Northeast*

State	Hay				Pasture					Total hay and pasture
	Alfalfa-grass	Red clover-grass	Other species	Total hay	Cropland	Improved	Other	Woodland	Total pasture	
					(000 acres)					
Maine	16	254	85	355	143	15	83	129	370	734
New Hampshire	18	91	31	140	62	7	40	110	219	363
Vermont	129	317	114	560	183	51	355	407	996	1,580
Massachusetts	30	95	22	147	80	12	61	86	239	398
Rhode Island	3	8	3	14	12	2	6	5	25	41
Connecticut	33	65	28	126	56	19	80	67	222	357
New York	1,177	1,251	322	2,750	991	223	2,079	816	4,109	6,993
New Jersey	74	58	16	148	88	18	66	19	191	352
Pennsylvania	790	1,087	95	1,972	686	281	1,238	573	2,778	4,865
Delaware	5	16	8	29	32	3	9	9	53	84
Maryland	91	200	43	334	226	129	212	111	678	1,038
West Virginia	110	345	88	543	360	319	1,286	910	2,875	3,428
Total	2,476	3,787	855	7,118	2,919	1,079	5,515	3,242	12,755	20,233

Source: USDC, 1966 (4).
Note: Nearest 1000 acres.

which alfalfa is seeded in the Northeast are, depending on location, timothy, smooth bromegrass, and orchardgrass. New York was the only state in the region that grew over a million acres of alfalfa for hay purposes. Pennsylvania was the next highest producer. Alfalfa hay produced in the Northeast in 1964 totaled 5.1 million tons.

Hay production on land unsuited for alfalfa and red clover depends upon birdsfoot trefoil in grass mixtures or on nitrogen-fertilized grass. Timothy is the most common grass sown with birdsfoot trefoil. Other grasses that associate well with this legume in mixtures are smooth bromegrass, late maturing strains of orchardgrass, and reed canarygrass. The four perennial grasses mentioned can also be grown alone with nitrogen fertilization for hay purposes. When nitrogen is used, general recommendations call for annual applications of at least 100 pounds of nitrogen per acre in conjunction with phosphorus and potassium. In the southern part of the region approximately 45,000 acres of annual lespedeza are grown for hay. Most of the lespedeza hay is produced in Maryland, but Delaware and West Virginia also share in its production. Other forage species account for approximately 1.4 million tons of hay annually in the region.

In addition to hay, corn and grass silage are the other harvested forage crops grown for feed in the Northeast. As indicated in Table 8.1, corn is the more important of the two silage crops grown. In 1964 approximately 13 million tons of corn silage were produced on an area slightly in excess of 1 million acres. Grass silage was produced on 360,000 acres the same year with approximately 1.9 million tons of grass ensiled.

PASTURE RESOURCES OF THE REGION

THE 12.8 MILLION ACRES of pastureland are divided into four classes: cropland pasture, improved pasture, other, and woodland pasture. Cropland pastures generally produce the most forage because they are confined to the more fertile, well-drained soils and are farmed in crop rotations. In 1964 the Northeast had approximately 2.9 million acres in cropland pastures. Forage species sown on cropland pastures are either perennials or summer and winter annuals. Orchardgrass, smooth bromegrass, reed canarygrass, and timothy are seeded with alfalfa or birdsfoot trefoil on these pastures. Ladino clover has been a favorite legume in the past, but because of its short life of two years it has lost favor with farmers. Once the legume disappears, the tall-growing grass pastures are fertilized with nitrogen to maintain production.

Sudangrass and sorghum-sudan hybrids are the summer annuals seeded on cropland pastures. Such small grains as rye, winter barley, wheat, and spring oats are also seeded on these pastures for fall forage and early spring grazing. Domestic ryegrass and field bromegrass can also be used for this purpose. Many cropland pasture acres are double cropped—summer annuals in season, followed by small grains, and vice versa. The summer and winter annuals generally supplement the semipermanent pastures seeded to the tall-growing legumes and grasses on cropland.

Improved (permanent) pastures cover approximately 1 million acres in the Northeast. They have been improved by liming and fertilization, renovation, or both. Basically these pastures are predominantly Kentucky bluegrass swards with wild white clover and a mixture of orchardgrass, timothy, and quackgrass. If they have been renovated, a long-lived legume such as birdsfoot trefoil and one or more of the tall-growing grasses also have been introduced. If the legumes have disappeared from the sward, it is top-dressed with nitrogen or a complete fertilizer is applied in spring and again in midsummer to maintain production.

By far the largest acreage of pasture, 5.5 million acres, in the region is classified as "other." For the most part, this is permanent pasture consisting of various proportions of one or more of the following species: Kentucky bluegrass, Canada bluegrass, redtop, bentgrass, wild white clover, and unpalatable weeds. Many of these pastures do not lend themselves to improvement because of shallowness of soil, poor internal drainage, rocks and rock outcrops, and steepness of slope that prevents use of farm machinery. Newly seeded pastures and hay aftermath may also fit into this pasture classification.

A sizable acreage of pasture, 3.2 million acres in the 12 states, is classified as woodland pasture. This acreage may include cutover woodland or pastures where brush and woody species have taken over. For the most part, woodland pastures furnish meager amounts of edible forage and browse.

Census data indicate that the Northeast ranks second to its timber resources in the production of forage and pasturage. Much of the land in the region cannot be utilized for cultivated crops because of natural limitations. If such land is to provide its owners an economic return, two common alternatives suggest themselves: (1) production of forage and pasturage in support of a livestock industry and (2) timber production. The former is often chosen because quicker returns can be realized, whereas the latter is a long-term investment.

IMPORTANCE OF LIVESTOCK

BECAUSE THE NATURAL FOOD for ruminant animals is forage, it is logical that livestock farming has become the dominant type of farming within the Northeast. Cattle are the most important heavy grass- and forage-consuming livestock. As indicated in Table 8.3 (4), dairy cattle are the most numerous of forage-consuming livestock. There are approximately 3.4 million milk cows in the region exclusive of young stock and bulls. Beef cattle and sheep total slightly in excess of 1 million head. In addition northeastern producers raise approximately 1.2 million hogs, 56 million chickens, 20.2 million broilers, and 6 million turkeys (Table 8.3). Hogs and poultry are heavy grain-consuming animals but do utilize some forage, depending upon the method of rearing.

Additional impetus to a livestock economy has been provided by the ready market for dairy and livestock products among the 55 million urban population (1970 census) who live in the large industrial cities within the region. The 7.5 million head of cattle, sheep, and hogs and the 82.3 million units of poultry on northeastern farms help meet this demand for livestock products.

FEED CONSUMPTION BY LIVESTOCK

ANNUAL FEED CONSUMPTION in feed units (equivalent in feeding value to one unit of corn per pound or ton) by livestock on northeastern farms is reported in Table 8.4 (1). Total feed units consumed from all sources was 26.8 million tons (1). This tonnage was divided among the feed sources as follows: concentrates, 12.5 million tons; harvested forage, 7.9 million tons; and pasturage, 6.4 million tons. On a regional basis, 54 percent of the feed units was derived from concentrates and 46 percent from forages. The latter included 25.1 percent from harvested forage and 20.9 percent from pasturage.

As indicated in Table 8.4, considerable variation exists in sources of feed units from state to state. In part this reflects the types of livestock enterprises of major importance in the various states. In Dela-

TABLE 8.3. *Livestock on farms in the Northeast*

State	All cattle	Milk cows	Beef cattle	Sheep	Hogs	Chickens	Broilers	Turkeys*
				(000)				
Maine	158	99	9	32	19	4,796	533	44
New Hampshire	82	53	2	4	14	1,831	215	78
Vermont	367	271	3	7	10	640	20	30
Massachusetts	127	91	3	10	118	2,806	607	332
Rhode Island	15	13	...	2	10	433	168	18
Connecticut	124	85	3	8	17	4,278	1,809	192
New York	1,959	1,338	49	126	89	11,973	427	452
New Jersey	153	125	5	13	160	7,188	251	299
Pennsylvania	1,789	949	101	192	452	17,637	3,014	1,949
Delaware	37	26	4	5	38	704	5,380	386
Maryland	416	226	45	28	187	1,983	5,833	223
West Virginia	483	122	174	209	73	1,771	1,919	2,083
Total	5,710	3,398	398	636	1,187	56,040	20,176	6,086

Source: USDC, 1966 (4).
Note: Nearest 1000.
* Breeding stock and raised stock.

TABLE 8.4. Feed units consumed by all classes of livestock in the Northeast as concentrates, harvested forage, and pasture, 1964

State	Concentrates	Harvested forage	Pasture	Total	Concentrates	Harvested forage	Pasture	Harvested forage and pasture
	(000 tons feed units)				(%)			
Maine	830	219	187	1,236	67.2	17.7	15.1	32.8
New Hampshire	224	118	89	431	52.0	27.4	20.6	48.0
Vermont	504	584	370	1,458	34.6	40.0	25.4	65.4
Massachusetts	436	205	147	788	55.3	26.0	18.7	44.7
Rhode Island	51	27	18	96	53.1	28.1	18.8	46.9
Connecticut	481	195	135	811	59.3	24.0	16.7	40.7
New York	3,168	3,040	2,051	8,260	38.4	36.8	24.8	61.6
New Jersey	701	269	187	1,157	60.6	23.2	16.2	39.4
Pennsylvania	3,567	2,338	2,018	7,923	45.0	29.5	25.5	55.0
Delaware	668	49	45	762	87.7	6.4	5.9	12.3
Maryland	1,354	483	452	2,289	59.2	21.1	19.7	40.8
West Virginia	561	332	693	1,586	35.4	20.9	43.7	64.6
Total	12,546	7,859	6,392	26,797	avg. 54.0	25.1	20.9	46.0

Source: Allen et al., 1970 (1).
Note: Figures do not exactly match those in Table 8.5 because of rounding.

ware, 87.7 percent of the state's feed units is derived from concentrates because of the heavy demand for grain by the broiler industry. Similarly, other states within the region with large poultry populations utilize high percentages of concentrates. Conversely, states with large cattle numbers (Vermont, New York, Pennsylvania, and West Virginia) obtain the highest percentage of feed units from harvested forage and pasturage. Vermont heads the list in this regard, followed by New York. This may be due in part to the long winters in those states, necessitating more barn-feeding. West Virginia derives 43.7 percent of its feed units from pasture, more than other states in the region.

Table 8.5 summarizes consumption of feed units in the northeastern states according to types of livestock (1). This is also illustrated in Figure 8.2. Dairy cattle consume the greatest tonnage of feed units of any livestock class in the area. Milk cows are fed 52 percent of all the feed units consumed in the region. The dairy industry utilized 64.7 percent of all the feed units fed in the Northeast for its milk cows and replacement stock. All classes of domestic ruminant livestock consumed 74 percent of all the feed units utilized in the Northeast compared with only 26 percent for the monogastric animals.

For the region as a whole, milk cow rations consisted of 36.9 percent concentrates, 44 percent harvested forage, and 19.1 percent pasturage in 1964; therefore forage consumption averaged 63.1 percent in contrast with 36.9 percent for concentrates. Replacement dairy cattle diets consisted of 16.6 percent concentrates and 83.4 percent forage.

TABLE 8.5.   *Feed units consumed by various livestock classes in the Northeast as concentrates, harvested forage, and pasture, 1964*

Livestock class	Concentrates	Harvested forage	Pasture	Total	Concentrates	Harvested forage	Pasture
	(000 tons feed units)				(%)		
Milk cows	5,141	6,142	2,657	13,940	36.9	44.0	19.1
Other dairy cattle	566	1,124	1,717	3,407	16.6	33.0	50.4
Cattle on feed	198	68	29	295	67.1	23.1	9.8
Other beef cattle	152	350	1,428	1,930	7.9	18.1	74.0
Sheep	25	19	204	248	10.1	7.7	82.2
Hogs	837	...	42	879	95.2	...	4.8
Hens and pullets	2,627	...	49	2,676	98.2	...	1.8
Chickens raised	612	...	28	640	95.6	...	4.4
Broilers	2,015	...	...	2,015	100.0	...	...
Turkeys	282	...	14	296	95.3	...	4.7
Horses and mules	100	154	220	474	21.1	32.5	46.4
Total	12,555	7,857	6,388	26,800	46.9	29.3	23.8

Source: Allen et al., 1970 (1).
Note: Figures do not exactly match those in Table 8.4 because of rounding.

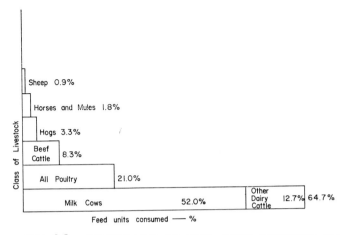

FIG. 8.2. PERCENTAGES OF FEED UNITS CONSUMED BY EACH LIVESTOCK CLASS IN THE NORTHEAST.

Sheep and beef cattle utilized the smallest percentage of concentrates in their rations and the highest percentage of forage. Concentrate consumption ranged from 7.9 percent for beef cattle to 10.1 percent for sheep, whereas forage consumption ranged from 89.9 percent for sheep to 92.1 percent for the beef cattle.

Among the monogastric animals, horses and mules consumed the least percentage of concentrates and the highest percentage of forage feed units. These animals received only 21.1 percent of their rations as concentrates and 78.9 percent as forages. As would be expected, poultry utilized the highest percentage of concentrates and the smallest percentage of forage in their rations. From 95.2 to 100 percent of their diets was made up of concentrates, with forages ranging from zero for broilers to less than 5 percent for other classes of poultry depending upon the method of rearing. Forage consumption by hogs measured in feed units was in the same range as poultry, less than 5 percent.

VALUE OF FORAGE
TO THE LIVESTOCK INDUSTRY

As INDICATED in Table 8.6, the 12 northeastern states produce approximately 14.3 million tons of forage feed units annually in support of the $2 billion livestock industry (1). This is shown graphically in Figure 8.3. Based on the price of corn, which ranged from $46.80

TABLE 8.6.   *Total forage feed units consumed by livestock value of forage consumed and livestock sales in the Northeast, 1964*

State	Forage feed units	Estimated forage value*	Livestock sales	Livestock feed units from forage
	(000 tons)	*(mil $)*		*(%)*
Maine	406	$ 24.0	$112.0	32.8
New Hampshire	207	12.2	38.9	48.0
Vermont	954	56.3	101.7	65.4
Massachusetts	352	20.8	82.7	44.7
Rhode Island	45	2.7	10.6	46.9
Connecticut	330	19.5	84.8	40.7
New York	5,091	249.5	601.9	61.6
New Jersey	456	23.8	107.4	39.4
Pennsylvania	4,356	216.1	578.1	55.0
Delaware	94	4.4	80.1	12.3
Maryland	935	43.8	182.4	40.8
West Virginia	1,025	51.7	69.3	64.6
Total	14,251	$724.8	$2,049.9	avg. 46.0

Source: U.S. agricultural census for crop year 1964, collected in 1965, published in 1966.

* Based on 1964 per ton corn prices, ranging from $46.80 to $59 per ton within the region.

per ton in the southern part of the region to $59 in the northern part during 1964, the forage feed units derived from hay, silage, and pasturage were valued at $724.8 million.

New York produced over 5 million tons of feed units derived from forage, Pennsylvania was next with over 4.3 million tons, and West Virginia followed with slightly in excess of 1 million tons. Feed production from forage approached 1 million tons of feed units in Vermont and Maryland.

The 14.3 million tons of feed units helped feed livestock in an industry that returned $2049.9 million to farmers in the Northeast

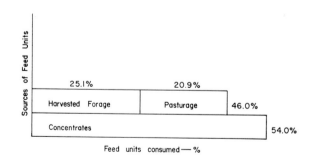

FIG. 8.3.   PERCENTAGES OF FEED UNITS FROM ALL SOURCES CONSUMED BY LIVESTOCK IN THE NORTHEAST.

during 1964. The forage produced in the Northeast was valued at 35.4 percent of livestock sales made within the region. Sales from ruminant animals and their products such as meat, milk, cream, and wool accounted for 68.1 percent of farm receipts in 1964. The greatest contribution to livestock receipts in the Northeast was made by the dairy industry. Milk and cream sales alone accounted for $1153 million. Sales of livestock products from monogastric animals in the form of meat and eggs approximated $653 million, only 31.9 percent of all 1964 livestock receipts. Since ruminant animals are heavy consumers of forage and pasturage in contrast with monogastric animals which are heavy grain consumers, this emphasizes the value of forage crops in the agricultural economy of the Northeast.

Preliminary 1969 data (3) do not indicate a drastic change in the agriculture of the Northeast over the previous five years. A slight reduction in cattle and sheep numbers has taken place in the region since 1964. On the other hand, the number of horses, ponies, and hogs has increased. Since the ruminant diet utilizes large amounts of forage crops and pasturage and nutrients from these feed crops generally can be produced cheaper than concentrates can be shipped into the region, forage crops and pastures can be expected to continue to receive major emphasis on northeastern livestock farms.

SUMMARY

Approximately 127.6 million acres of land are in the Northeast but only 41.2 million acres of this land are in farms. Of this acreage, 51.9 percent is committed to forage and pasture production in support of the 6.3 million head of ruminant livestock in the region. In addition, 1.2 million hogs and 82 million chickens, broilers, and turkeys are raised. All classes of livestock consume 26.8 million tons of feed units annually. These feed units are derived as follows: 54 percent from concentrates, 25.1 percent from harvested forage, and 20.9 percent from pasturage (Fig. 8.3). Dairy cattle utilize 64.7 percent of all the feed units fed in the Northeast. All classes of ruminant livestock consume 74 percent of all the feed units fed in the region, compared with only 26 percent for the monogastric animals.

The 12 northeastern states produce approximately 14.3 million tons of forage feed units annually in support of its $2 billion livestock industry. The forage feed units derived from the hay, silage, green chop, and pasturage grown in the region are valued at $724.8 million based on the price of corn. Hence, the 21.4 million acres of north-eastern land devoted to forage crops and pastures provide a sound economic basis for its livestock industry. In addition, since approximately 95 percent of this acreage is kept in sod crops, such fringe benefits can be expected as conservation of soil and water and improvement of soil structure, soil productivity, and groundwater quality.

REFERENCES

1. Allen, G. C., E. F. Hodges, and M. Devers. 1970. National and state livestock-feed relationships. USDA Stat. Bull. 446, suppl.

2. Sprague, H. B. 1959. Grasslands. Am. Assoc. Adv. Sci. Publ. 53., Washington, D.C.

3. USDA. 1970. Agricultural statistics.

4. USDC. 1966. 1964 U.S. census of agriculture. Bureau of Census.

5. Washko, J. B. 1960. Early history of grassland management. Agron. Dept. Seminar, The Pennsylvania State University.

9

GRASSLANDS AND FORAGE IN THE AGRICULTURAL ECONOMY OF THE MIDDLE STATES REGION

S . H . D O B S O N

THE MIDDLE STATES (or Upper South) region lies across the 37th parallel, extending at places three parallels on either side. It begins at the Atlantic Seaboard and extends to the western boundaries of Missouri and Arkansas. The soil region described in the 1957 Yearbook of Agriculture as the east-central uplands (Fig. 9.1) covers the most extensive land mass (5). However, the subgroups are quite different and are significant. They include the Piedmont Plateau of North Carolina, Virginia, and Maryland; the Blue Ridge in North Carolina, Virginia, and Tennessee; the Appalachian Valley in Virginia and Tennessee; the Allegheny-Cumberland highlands in Tennessee, Kentucky, and West Virginia; the bluegrass areas in Kentucky and Tennessee; the Ozarks in Arkansas and Missouri; and the Cherokee-Parsons soil area of Missouri. Some of these are broken down into sub subgroups.

The Middle States region also includes the coastal plain sections of Maryland, Virigina, and North Carolina; the Mississippi delta sections of Missouri and Arkansas; and the midland feed region in Missouri.

Elevations vary from over 6000 feet to sea level. Rainfall varies from 36 inches to 55 inches with most of the area within the 40- to 50-inch rainfall zone. The frost-free days range from 128 to 269, with 190 days representing a major block of the forage lands. Relatively mild periods after frost in the fall and before the last frost in the spring permit considerable growth of cool-season species such as tall fescue.

113

FIG. 9.1.   THE MIDDLE STATES REGION.

POTENTIAL FOR GROWTH

THE 1948 USDA Yearbook of Agriculture gives the point of departure
for this discussion. Lovvorn has described the area as a wide and
versatile empire (4). Several authors have emphasized the increasing
efforts in grassland research, the development of a technological background, and the fact that this information was being put to use by
more and more animal farmers. With some ups and downs, this trend
continues, but the vast potential is unexplored. The favorable amount
of rain falling on millions of acres of the mid-South, the unfertilized
pastures and only partially fertilized grasslands, and the many acres
of unused area remain to be exploited. The increasing population
plus the increasing per capita demand for red meat seem to indicate
that there is an opportunity to capitalize on these resources. Since
the yearbook was written, the area has increased its animal numbers
by about 150 percent. North Carolina, for example, has doubled its
agricultural income from livestock and livestock products since 1958

and had one of the largest percentage increases in beef cattle numbers in the United States in 1969.

In the final analysis, how far the region should go in the direction of grasslands will be determined by the profit potential and whether grassland can be competitive with other uses for the land, labor, capital, and management resources of the region.

Production Experiences. Decker (2) and associates in Maryland have averaged 572 pounds of beef per acre for seven years on a combination of Midland bermudagrass and sod-seeded cereals. This gives a gross of $150–$175 per acre. But considering the minimal labor, machinery, and harvesting costs, we can quickly sense the possibilities. When converting these mixtures to hay, these same workers achieved as high as 10 tons per acre, with returns above seeding and fertilizer costs of over $200 per acre.

Blazer (1) and co-workers in Virginia have conducted many grazing trials where animals were utilized in measuring forage yields and treatment differences. They clearly show the problems of seasonality of midsouthern forages. The variability is due to changes in temperature and irregular rainfall, which result in ups and downs in forage yield and quality. Factors such as animal types, stocking rates, and grazing species add variability and complications for the grasslander to consider and solve. However, forage yields of 3–5 tons per acre and animal gains of over 400 pounds per acre are obtainable in Viriginia, in many cases from soils that do not lend themselves to cultivation.

West Virginia has a good example of forage-meat potential on sloping pastures. Here, aerial applications of fertilizers and herbicides were used and performance measured. Increases in animal numbers of 30–40 percent were noted with an additional 10–40 percent increase in calf or lamb gains. Virginia and North Carolina show that minimum applications of phosphate and lime can double yields, with 100 pounds of P_2O_5 applied every 3 years and 0.5 ton of lime applied once in 12 years. The cost of total digestible nutrients based on these yields, even with the cost of aerial fertilizer application, is quite reasonable and is competitive in price with forage produced from many other systems in the region.

Kentucky has over 9 million acres of forage and grasslands, and probably has the most intensive research efforts in the area. Research workers in Kentucky and Tennessee have pioneered the efforts in pasture renovation, which in this area consists of the addition of a legume, primarily white clover, by disking or otherwise reducing the stands of old, predominantly fescue sods. Several million acres have been renovated, doubling yields of forage per acre and increasing per animal performance. This practice is spreading into North Carolina and other adjacent states.

Missouri, which has the largest animal population in the area, has long been a leader in forage research and forage improvement programs. One of the spectacular stories of grassland salesmanship is the Howard County Grassland Improvement Program. Topdressed pastures and hayland have gone from 5000 acres in 1961 to 37,000 acres in 1969. Initially, demonstrations on a few acres showed that yields could be doubled and tripled by fertilizer applications. As a result, farmers could feed and sell more and larger animals from the same acreage, pay their bills, and show a profit. Similar research stories, educational programs, and production trends exist in each of the states.

TYPES OF FORAGE

THE MIDDLE STATES region might be classed as the ladino clover belt, with tall fescue and orchardgrass the primary companion grasses. Kentucky bluegrass volunteers freely in each of the states and in many locations is still the most desirable species. Red clover enjoys considerable popularity with a regional breeding and improvement program in Kentucky. Lespedezas are still quite common but decreasingly so.

Production of alfalfa, once quite important in the area, had declined but is now bounding back. The development of weevil-tolerant varieties and the buildup of weevil parasites, plus the fact that no real satisfactory substitute for alfalfa has emerged, have revived farm interest in the crop. Yields of 5–8 tons per acre are attainable. The soil-holding and soil-building characteristics, perennial nature, and high yield of protein and generally high quality of the forage make it very desirable for the region. However, corn silage is more nearly mechanized, and on good corn land, stored feed can be produced cheaper from corn than any other source. It would appear that a combination of alfalfa and corn is ideal from the standpoint of both production and feeding.

THE GRASSLAND RESOURCE IS OF GREAT IMPORTANCE TO OUR NATIONAL LIFE.

Coastal bermudagrass–tall fescue pasture is grazed by Hereford cattle near Waterboro, S.C. Below: High-quality ladino clover–reed canarygrass forage for high-producing Holstein cows is grown on Massachusetts land once considered too wet to use. *SCS photos*

ABOUT 43 PERCENT OF OUR GRAZING LAND RESOURCES ARE PRIVATELY OWNED.

Holstein dairy cattle graze on pasture of well-fertilized orchardgrass, redtop, red clover, and alfalfa near Bowling Green, Ky. Below: Good management is exemplified by the bermudagrass pasture on the right where grazing was deferred during the growing season in Collin County, Tex.; on the left is continuously grazed pasture. *SCS photos*

LIVESTOCK PRODUCTION WOULD BE DIFFICULT AND COSTLY WITHOUT FORAGE FEEDS.

Coastal bermudagrass provides feed for this Arkansas beef operation. Below: Ruminant livestock make very effective use of dry rangeland such as this near Canyon, Tex. *SCS photos*

RUMINANTS CAN CONVERT PLANT MATERIAL THAT IS UNUSABLE BY OTHER ANIMALS INTO FORMS HIGHLY PALATABLE TO MAN.

High-quality Angus cattle graze salt marsh rangeland near Cameron, La. The dominant grass is seashore paspalum. Below: Hereford cattle graze a grass-legume mountain meadow in Elko County, Nev. *SCS photos*

GRASS COVER GREATLY BENEFITS OUR WATER RESOURCES.

Watershed protection is provided by bahiagrass-lespedeza pasture in Tallahatchie County, Miss.; a combination livestock and fish pond is in the background. *SCS photo*

FORAGE WOULD HAVE LITTLE VALUE SAVE FOR LIVESTOCK.

The zebu (Brahman) type of cattle are widely grown in tropical and subtropical regions around the world. This region in Texas is unsuited for cropping but provides forage to be transformed into protein for man. *SCS photo*

PRESERVATION OF FORAGE ALLOWS FEED HARVEST AND STORAGE TO BE
HANDLED RAPIDLY AND ECONOMICALLY FOR RETENTION OF MAXIMUM
FEEDING VALUE.

The windrower is a valuable attachment to the mower for hay-curing
and field-wilting of alfalfa for silage. *SCS photo*. Below: Hay may be
dried by forcing air through bales in special wagons. Hay is baled in
the field when partially cured and then mechanically loaded in a ran-
dom bale arrangement.

Contented Jersey cows graze on improved pasture in Chittenden
County, Vt. Below: River bottom pasture is grazed in Chester County,
Pa. *SCS photos*

GREAT POTENTIAL EXISTS FOR EXPANSION IN PRODUCTION OF LIVESTOCK AND LIVESTOCK PRODUCTS IN THE MIDDLE STATES.

Ladino clover–tall fescue provides good pasture in North Carolina. Ladino is widely used in the area but needs periodic renovation to maintain a satisfactory stand. Below: Yearling steers graze orchardgrass pasture in Cabell County, W. Va. Heavily grazed for two months, this pasture has maintained growth with good management. *SCS photo*

BEEF CATTLE REPRESENT THE LARGEST MARKET FOR FORAGES IN THE NORTH CENTRAL REGION.

Improved pasture of reed canarygrass offers good grazing for a beef cow-calf enterprise in Menominee County, Mich. *SCS photo*

TREMENDOUS OPPORTUNITIES EXIST IN THE HUMID SOUTH FOR INCREASING PRODUCTION PER ACRE.

Several tropical grasses not usually seen elsewhere are important in Florida; Brahman cattle graze Pangola digitgrass pasture in Hernando County. *SCS photo*

THE GREAT PLAINS STATES ARE UNIQUELY SUITED TO LIVSTOCK PRODUCTION ENTERPRISES.

Broad, dry valleys and gently rolling hills typify the sandhill region of Nebraska. These are highly productive cow-calf rangelands. Below: Stock water dams make possible the uniform utilization of the productive northern Great Plains grasslands in South Dakota. *SCS photos*

THE VAST RESOURCES OF RANGELAND ARE CONVERTED BY RUMINANTS INTO
PROTEIN FOR HUMAN CONSUMPTION.

Cattle on seeded intermediate wheatgrass range in Utah. The area
was reclaimed from thick pinyon-juniper growth three years prior to
this picture. Below: Sheep graze on spring-fall bunchgrass in Wasco
County, Oreg. Due to careful management, this dry range is in good
condition for its class. *SCS photos*

IN MANY AREAS OF THE WEST, PUBLIC DOMAIN LAND IS A VITAL PART OF THE TOTAL LIVESTOCK OPERATION.

Desert federal rangeland is in good condition here where native desert grasses coexist with Joshua trees. Below: Semiarid federal rangelands in a typical valley of the western Great Basin area provide needed forage for the herds of nearby ranchers. *BLM photos*

Cattle graze in Custer National Forest on national grassland ranges administered by the U.S. Forest Service. Below: Rolling grassland slopes provide good cattle grazing in the Helena National Forest in Montana. *U.S. Forest Service photos*

LIVESTOCK GRAZING AND WILDLIFE MANAGEMENT CAN BE COMPLEMENTARY
ENTERPRISES.

Graceful pronghorn antelope share this grazing land with man's livestock. Below: A stock watering pond serves livestock that feed on this
dry range in Musselshell County, Mont.; it also is well used by wildlife. *SCS photo*

Recycling cattle feedlot effluent for irrigation of grassland is an effective method of preventing pollution from animal wastes. Below: Parallel terraces are particularly effective in soil and water conservation where erosion is otherwise serious. *SCS photos*

Beauty, safety, and erosion control are provided by bahiagrass along a roadside in Treutlan County, Ga. Below: Grass protects banks and adds beauty to a recreational fish pond in Knox County, Ind. *SCS photos*

A skeleton planting of beachgrass will fill in solidly to stabilize sand dunes in Ottawa County, Mich. Below: Hydraulic seeding of grass on a watershed project structure provides for flood prevention. *SCS photos*

TABLE 9.1. *Changes in acres of forages in the Middle States, 1949–64*

State	Hay			Pasture			Silage		
	1949	1959	1964	1949	1959	1964	1949	1959	1964
Maryland	413,956	408,493	368,003	961,733	781,666	768,363	11,789	33,777	131,618
Virginia	1,173,036	1,094,202	1,006,765	5,299,045	4,723,241	4,575,190	99,985	110,573	202,745
West Virginia	261,644	601,401	567,379	4,319,092	3,222,411	2,875,249	20,613	28,532	39,580
North Carolina	965,599	612,034	530,508	3,212,637	2,738,221	2,716,674	63,619	88,700	133,225
Kentucky	1,676,388	1,471,039	1,597,501	8,570,544	7,765,993	7,829,352	69,100	59,502	99,479
Tennessee	1,666,667	1,194,588	1,260,006	6,415,280	6,445,560	6,605,456	82,174	62,637	102,682
Missouri	3,577,164	2,471,427	2,964,001	16,944,928	15,549,645	15,476,261	157,752	228,681	312,745
Arkansas	1,033,411	636,967	699,891	7,301,407	6,895,280	7,135,321	38,567	32,055	35,535
Total	10,767,865	8,490,151	8,994,054	53,024,666	48,122,017	47,981,866	543,599	644,457	1,057,609

TABLE 9.2. *Total of all forages by state, Middle States*

State	1949	1959	1964
Maryland	1,387,478	1,223,937	1,267,984
Virginia	6,572,066	5,928,018	5,784,700
West Virginia	4,601,349	3,852,344	3,482,208
North Carolina	4,241,855	3,428,955	3,380,407
Kentucky	10,316,032	9,296,534	9,526,332
Tennessee	8,164,121	7,702,755	7,968,144
Missouri	20,679,844	18,149,753	18,753,007
Arkansas	8,373,385	7,564,311	7,870,747
Total	64,336,130	57,146,607	58,033,529

ECONOMIC ANALYSIS

IN UNDERSTANDING the economics of an enterprise, Jacobs (3) has made some really valid points in his attempts to bring about economic analysis of grasslands and forages in Missouri. There is a general awareness that data on forage yields and gross animal gains on any particular pasture or forage are not enough. The key is how to combine these to bring about year-end maximum returns (for all the acres and capital used) to pay labor, fertilizer, and other bills and realize a reasonable return for management.

North Carolina has set up a team of specialists, primarily in agronomy, animal husbandry, and farm management, to work with beef farmers along these lines. One farmer in eastern North Carolina, who in 1966 analyzed his beef cattle business, decided his enterprise was not paying enough for the use of his land. Corn and soybeans were more profitable. He decided to do better or get out of the beef business. In 1966, after paying all the cost of labor, interest, maintenance of farm roads and ditches, machinery depreciation, etc., he had only $12.50 return per acre of land devoted to pastures. By 1969, this had increased to $44.62 per acre. In addition, he had doubled his charges for beef cattle management during this period. He has expanded his cattle business on these same acres and plans to use additional acres as his animal numbers build up. This farmer is selling over 600 pounds of beef from every acre devoted to grassland and forages on the farm. Again, this can be misleading because he utilized field gleanings and winter cover from the corn and soybean land and from diverted acres on the farm. However, this is part of the "enterprise mix" that looks so good to the better farm managers of the region. True, his return per acre from corn and soybeans would be greater if all the acres were planted to these crops, but the net to the farm has increased substantially by expanding beef cattle.

In the Piedmont the land is rolling and erosive and only a small percentage is capable of growing corn at a profit. Here beef farmers are netting in excess of $30 per acre for use of the land. Many of these are part-time farmers who own amounts of cleared land varying from 10 to 200 acres. They can easily manage these animals as a side enterprise and increase the family income substantially.

All this seems to indicate that large farmers can consider grassland and forages as capable of being (in many cases they already are) quite profitable, when fed to beef animals. It is likely that beef will increase more as an "enterprise mix" rather than entire farms being devoted to its production. In addition, the part-time farmer will probably contribute substantially to the movement toward grassland and forages utilized by beef.

Dairying as a farm enterprise yields very good economic returns. In North Carolina during 1969 the top 20 percent of dairymen in the farm business record program averaged over $20,000 per year return to management. The top 10 percent went to nearly $30,000 and some went over $40,000 (according to 126 records). Certainly this will be competitive with many enterprises for the best managerial talent.

Dairy farmers in the entire region are moving more and more to the use of silage, primarily corn. Blight and some other factors might shift this trend toward sorghums and more haylage from alfalfa and small grains. As an example of the trends in dairy feeding, consider the Dairy Herd Improvement Association herds in North Carolina. In 15 years the consumption per cow has changed as follows: silage increased from 7500 to 18,400 pounds; hay decreased from 2900 to 1000 pounds; days on pasture decreased from 215 to 135; grain fed increased from 2900 to 5300 pounds. Production has increased from 8466 to 12,532 pounds during this period.

SUMMARY

THE FACT that grassland is increasing in importance is attested to by *The Southern Planter,* a magazine that has added a pasture and forage supplement. Other encouraging developments include the organization of a state grassland council in Kentucky with a forage symposium in 1970; the very fine 20-year summary of forage research at the Virginia Forage Research Station, Middleburg, Virginia; the development of aerial application of fertilizers and herbicides to mountain pastures; the state forage analysis programs; and the fact that some leading farmers, who once questioned the economic returns for beef, offer some very positive and profitable answers.

REFERENCES

1. Blazer, R. E. 1969. Managing forages for animal production. Va. Polytech. Inst. Res. Bull. 45.

2. Decker, A. M. 1969. Midland bermudagrass forage production. Univ. Md. Agr. Exp. Sta. Bull. 484.

3. Jacobs, V. E. Pasture value is not gross yield. *Better Crops with Plant Food* 54:24–26.

4. Lovvorn, R. L. 1948. A wide and versatile empire. *In* Grass, pp. 455–58. USDA Yearbook Agr.

5. USDA. 1957. Soil Yearbook Agr.

10

CONTRIBUTION OF FORAGES
TO AGRICULTURAL INCOME
IN THE HUMID SOUTH

DAVID A. MAYS

ANYONE who travels through the South can see that grasslands are tremendously important to the region. Their value goes far beyond providing feed for livestock and includes such benefits as soil stabilization and improvement, runoff reduction, and lessening of siltation in waterways. Perhaps of equal importance is the aesthetic value of the beautiful green-carpeted fields seen throughout the region. Placing a realistic value on these grasslands in terms of their monetary contribution to the region's economy is quite difficult.

This chapter presents the results of an extensive survey of available information on the extent and value of southern grasslands compared with certain other facets of the region's agricultural economy. Additionally, an attempt has been made to establish long-term trends in land use, cattle production, sales of livestock products, and current fertilization practices in order to make some estimate of future economic potential for the grassland segment of agriculture. The primary published sources of information used in the study were the 1964 census of agriculture and the 1967 conservation needs inventory for each state of the region. The soil conservation services of each state and the agronomy, animal science, and, in some cases, agricultural economics departments of the land-grant universities were also consulted for help in estimating crop yields and making projections of the future trends in forage and livestock production in the region.

The humid South in this study includes the states of South Carolina, Georgia, Florida, Alabama, Mississippi, and Louisiana, plus southern Arkansas and eastern Texas (Fig. 10.1). Southern Arkansas was considered to be the 26 counties south of or astride a line from Memphis, Tenn., to the Texas-Oklahoma border. Humid eastern

121

FIG. 10.1.   THE HUMID SOUTH REGION (DOTTED AREA).

Texas was defined by the Texas soil conservation service as being the 57 counties lying east of precipitation-evaporation index line 64.

ANNUAL FORAGES

THE TERM *forage* encompasses a broad spectrum of annual and perennial crops of varying and constantly changing importance in the humid South. Annual forages include corn and sorghum grown for green chop or more often preserved as silage. Silage crops are usually found only on the larger farms and are fertilized and managed to produce the maximum tonnage of total digestible nutrients per acre. Yields of 10–15 tons of green weight per acre are common, and 25- to 30-ton yields are sometimes achieved. In most of the South only Coastal bermudagrass will equal a well-managed corn or sorghum silage crop in the production of large tonnages of forage for ruminants.

Sudangrass, pearlmillet, and the newer sorghum-sudangrass hybrids are often planted for summer grazing or for green chop. These grasses are used less often for silage because they are lower yielding than forage sorghums and are not favored for hay because of the difficulty in curing the heavy stems. Summer grazing crops are more likely to be planted for dairy cattle than for beef cattle because the high nutrient requirements of lactating cows are not adequately met by most of the perennial grazing crops presently grown in the South. Annual grasses are sometimes planted as emergency crops, but none is drought tolerant. They require good soil moisture levels for rapid growth and will provide adequate dry-weather grazing only if the forage has been allowed to grow and accumulate during previous periods of favorable moisture.

Historically, annual lespedeza occupied large acreages in the South. Lespedeza produces top quality hay and makes excellent pasturage for fattening calves prior to fall sale. However, it has lost favor because of low yielding capacity and is now most commonly found growing in association with a cool-season grass, where it furnishes some July and August grazing while in a period of low production.

Winter annual forages include ryegrass, the small grains (wheat, oats, barley, and rye), and several legumes. When these crops are planted early in the fall and good moisture prevails, they furnish much late fall and winter grazing particularly near the Gulf Coast. However, because of colder temperatures, production may be limited in the upper South.

Small grains in combination with annual clover and ryegrass produce the most rapid steer gains of any pastures grown in the South. Average gains often reach 2.2–2.5 pounds per day. Cattle may be removed from small-grain pasture in late winter so that a silage, hay, or grain crop can be harvested; but the pasturage is more often grazed out and followed by a summer crop.

Small grains have been sod-seeded in summer-growing perennial grasses such as bahiagrass or bermudagrass to extend the grazing season. Some research workers have reported the production of an extra 150–200 pounds of beef per acre by this method. In the lower South this practice shows less advantage because of the late fall competition from perennial grasses and the resulting short grazing season on rye or oats.

Winter annual legumes such as crimson, ball, and arrowleaf clover; hairy vetch; and caley peas may be overseeded on bermudagrass or bahiagrass sod or planted on prepared seedbeds either alone or in association with ryegrass and/or small grains. With proper management to ensure a seed crop, crimson and ball clover need not be replanted annually.

PERENNIAL FORAGES

PERENNIAL FORAGE GRASSES are generally classified as being cool-season or warm-season species. Tall fescue is the most widely adapted of the cool-season grasses and is the principal forage grass grown on many farms in the northern part of the Gulf states. It makes excellent spring and fall growth, but growth and carrying capacity are very limited in July and August. In the lower South tall fescue is useful for fall and early spring grazing, but summer production is minimal. Orchardgrass is adapted in the northern part of the region but is not widely grown. Kentucky bluegrass usually occurs only as volunteer patches in sods of other species.

Adapted warm-season grasses provide most of the forage for the cattle industry in much of the South. Coastal bermudagrass occupies some 4 million acres in Texas and 2 million acres in the other southern states. Its use is restricted to the southern two-thirds of the Gulf states and Georgia, plus the coastal plain of South Carolina, but it is adapted to all areas of the region except for the highest elevations in South Carolina and Georgia. Coastal bermudagrass is not usually grown in southern Florida since it is not as productive there as certain other species.

Midland bermudagrass is adapted to much the same areas as Coastal but will survive colder winters, while the new higher quality Coastcross 1 is winter-hardy only in the southern part of the bermudagrass belt. All the hybrid bermudagrasses are excellent for hay and make good early summer grazing, but from June onward animal gains on pasture are generally low. Common bermudagrass is less productive than the hybrids and produces lower quality forage; however, when properly fertilized and managed, it is quite productive. It is likely that the acreage of common bermudagrass considerably exceeds that of the hybrids, but much of it receives little or no fertilizer or management and its potential contribution is not fully realized.

Bahiagrass is the most important grass species in the southern half of the Gulf Coast states and much of Florida. It is more commonly used for grazing since efficient hay harvesting is difficult because much of the leaf growth is concentrated close to the soil surface. Argentine bahiagrass is used for hay to some extent along the Gulf Coast of several states. At fertilization rates up to 200 pounds of nitrogen per acre, bahiagrass yields equal those of Coastal bermudagrass, but bahiagrass generally will not respond as well as Coastal to extremely high nitrogen rates.

Although johnsongrass is generally considered to be a weed in much of the region, it fills an important spot in the forage picture in the Black Belt of Alabama and Mississippi where it is widely used for grazing and particularly for hay. Care must be taken to avoid overgrazing johnsongrass, or adequate stands cannot be maintained. Dallisgrass is important for grazing in much of the area and occupies more acreage than Coastal bermudagrass in some states.

Several tropical grasses not usually seen elsewhere are important in Florida. Pangola digitgrass is a very high-yielding upright grass used for grazing, hay, and silage in central and southern Florida. Paragrass and saint augustinegrass are also often found in Florida pastures. Other low-yielding warm-season grasses such as carpetgrass and centipedegrass occur to some extent in many native pastures but probably make a minimum contribution to the economy.

The lower South is fortunate in having a number of perennial grasses that are widely adapted throughout the region. Unfortunately, however, most warm-season perennial grasses are generally low in quality. The average daily gains that can be achieved by grazing animals are considerably less than those achieved on bluegrass or orchardgrass in more northerly regions of the country.

Perennial legumes are of less importance in the South than elsewhere in the United States. However, several varieties of white and ladino clover are widely grown on heavier soils or bottomlands, usually in pasture mixtures with tall fescue or dallisgrass. Grass-clover pastures are generally considered to be superior to straight grass pastures because of better animal performance and lower requirement for commercial nitrogen, but many farmers do not exercise the management necessary to maintain adequate clover populations. When warm-season perennials such as bermudagrass and bahiagrass are fertilized for maximum production, they are too competitive for clovers to be retained in the stand.

Alfalfa and red clover are of little importance anywhere in the region, with most states reporting only a few thousand acres even though both are well-adapted on many soils in the northern half of these states.

Sericea lespedeza is perhaps the most widely grown perennial legume on upland soils of the region. It is able to establish itself and persist on many sites where conditions are too poor for many other forage species. The usefulness of sericea has been limited by its high tannin content and stemminess which lower palatability. New cultivars such as Serala, which are being bred for improved palatability, will be likely to contribute to a significant increase in acreage of this useful legume.

A number of tropical legumes including hairy indigo, tropical kudzu, and *Stylosanthes humilis* have some usefulness in southern Florida where temperate zone legumes are not grown.

GRASSLAND ACREAGES AS RELATED TO
LAND CAPABILITY CLASS

THE SOIL CONSERVATION SERVICE has divided all farmland into land-capability classes ranging from I to VIII (1). These classes are based on the capability of the land to produce cultivated crops and pasture over a long period without soil deterioration. Factors affecting classification are erosion hazard (slope), wetness, rooting limitations, and climate. Class I lands are suitable for any agricultural use, while Class VIII lands have no commercial value.

Table 10.1 shows how the ratio of cropland to grassland acreage changes with land capability class in the six states for which data

TABLE 10.1   *Percentages of nonforested farmland in cropland and grassland by state and land capability class*

Land capability class	Land use	South Carolina	Georgia	Florida	Alabama	Mississippi	East Texas
I	Cropland*	94.6	85.3	83.3	74.2	87.6	27.7
	Grassland†	5.4	14.7	16.7	25.8	12.4	77.3
II	Cropland	79.3	75.1	62.5	57.3	53.3	28.3
	Grassland	20.7	24.9	37.5	42.7	46.7	71.7
III	Cropland	71.5	57.4	57.7	43.8	62.8	16.2
	Grassland	28.5	42.6	42.3	56.2	37.2	83.7
IV	Cropland	72.0	44.2	30.0	28.5	46.2	8.0
	Grassland	28.0	55.8	70.0	71.5	53.8	92.0
V	Cropland	33.9	30.1	40.9	11.5	79.7	2.2
	Grassland	66.1	69.3	59.1	88.5	20.3	97.8
VI	Cropland	45.3	29.9	69.0	17.9	17.9	1.9
	Grassland	54.7	70.1	31.0	82.1	82.1	98.1
VII	Cropland	40.5	36.2	31.6	14.2	12.0	0
	Grassland	59.5	63.8	68.4	85.8	88.0	100

Source: USDA, 1967 (3).
* Includes row and close-grown field crops, fruits, and vegetables.
† Includes permanent and rotation pasture, hay, and range.

were available (3). These data cover only nonforested farmland and exclude land in farmsteads and other noncrop uses. For inventory purposes, all land with 10 percent or more of trees and brush was considered as woodland even though much of it is also grazed. In all states except Texas, more than 70 percent of Class I land is used for row crops, small grains, fruits, or vegetables. As land becomes less favorable for tilled crops, proportionately more is used for sod crops.

The relationships between land-capability class and land use vary from state to state. This is somewhat dependent on which limiting factors are responsible for class designation and on the type of crops grown in different areas. For instance, soil wetness is a determining factor in classification, but wetness might not preclude use of the land for rice. Likewise, steep land might be suitable for sodded orchard use even though it fell in capability Classes VI or VII.

From the viewpoint of grassland importance, the significance of the data in Table 10.1 is that 70 percent or more of the region's Class IV to Class VII farmland and a significant amount of Classes II and III land is in grass. It is extremely unlikely that long-term productivity could be maintained on much of this land under the intensive cultivation needed for row crop production. However, new sod planting techniques may change this picture somewhat.

With proper forage species selection and management, grass swards can be almost as productive on many of these more difficult sites as on Class I to Class III land. Because of the wide diversity of species (ranging from drought-resistant Coastal bermudagrass to tall fescue or dallisgrass) that will grow on quite wet lands, few areas exist which cannot become productive grasslands.

Machinery operation presents fewer problems on grassland than on tilled acreage. Slope, soil wetness, rock outcrops, and other factors may severely limit tilled crop farming operations that often must be done frequently and on time. On the other hand, sod crops require infrequent tillage, and much leeway exists in the timing of necessary fertilization and pasture clipping operations.

TOTAL ACREAGE AND ECONOMIC VALUE OF FORAGES

THE ACREAGE and value of major categories of forage and field crops are shown in Table 10.2 (4). Perennial hay and silage crops include grasses and legume-grass combinations harvested for hay or silage. In several states of the region Coastal bermudagrass acreage was not

TABLE 10.2. *Acreage and value of forage and field crops in selected southern states*

State	Perennial hay and silage crops		Annual hay and silage crops		Pasture		Other major field crops	
	(000 acres)	(mil $)	(000 acres)	(mil $)	(000 acres)	(mil $)/§	(000 acres)	(mil $)
South Carolina*	166	$ 8.6	176	$10.0	1,874	$ 23.9	1,907	$ 233.1
Georgia†	817	40.3	264	25.8	5,418	64.0	3,068	309.8
Florida*	127	8.2	24	2.0	10,024	28.4	730	100.9
Alabama‡	436	16.7	133	8.1	6,696	84.8	2,098	145.4
Mississippi*	570	19.0	191	12.9	8,854	112.0	3,546	469.7
Southern Arkansas*	127	5.5	19	1.0	1,975	41.7	1,699	178.9
Louisiana*	423	15.5	71	4.1	5,461	83.2	3,244	303.8
East Texas*	666	24.1	145	6.1	13,243	304.6	1,383	166.6
Total	3,332	$137.9	1,023	$70.0	53,545	$742.6	17,675	$1,858.2

Source: USDC, 1966 (4).
* 1968 data. Supplied by Ext. Agron. Dept., Univ. Ga.
† 1969 data. Supplied by Agron. Dept., Auburn Univ.
‡ Estimated from Auburn University values of $4.69 per acre for woodland pasture and $18.75 for all other pasture except in Florida where values of $16.50 for cropland pasture, $1.32 for woodland pasture, and $6.60 for other pasture; in Arkansas where values of $40.00 for cropland pasture, $10.00 for woodland pasture, and $20.00 for other pasture; and in East Texas where values of $38.00 for cropland pasture, $6.00 for woodland pasture, and $30.00 for other pasture were supplied by the respective universities.

reported separately, although it is an important hay crop everywhere except in Arkansas and southern Florida. This probably resulted in underestimating the total value of hay since well-managed Coastal is higher yielding than other hay crops commonly grown. Less than 15,000 acres are devoted to hay crop silage in any of the southern states.

The annual hays and silages contribute about half as much to the agricultural economy as perennial harvested forages. About 35 percent of the annual crop acreage was devoted to corn and sorghum for silage. These crops were valued at $43 million in 1964. Much greater acreage was used for other annual crops such as small grains harvested for hay or silage and the millets, sudangrasses, and sudangrass crosses; but the total value was significantly less than for corn and sorghum silage because of the relatively low per acre yields and lower feeding value of the stored crop.

It is extremely difficult to place a dollar value on pasture production. Productivity varies tremendously from cropland pasture to woodland pasture and in response to management level. In addition, the actual recovered value depends on the extent of efficient utilization of available herbage by ruminant animals.

The agricultural census does not place a dollar value on pasture. Agronomists at Auburn University estimated that under present low levels of management, woodland pasture yields 0.25 ton of dry forage per acre worth $4.69, while open pasture yields approximately 1 ton and is worth $18.75 per acre. Since estimates were not received from all the Gulf Coast states, the Auburn figures were used in estimating the value for all pasture in the region except for Florida, Arkansas, and Texas where different estimates were supplied by university agronomists. Estimated pasture values were lower for Florida and higher for Arkansas and Texas. The southern region has more than 44 million acres of grazing land, of which about 40 percent is woodland pasture. Using the values indicated above, this pasturage has an estimated total annual value of $742.6 million. When this is added to the annual and perennial hay and silage crops worth $70.0 million and $137.9 million respectively, the total yearly forage output is worth some $950.5 million. This represents an average gross return of $19.56 per acre on the 48.6 million acres used for forage production.

In contrast to this forage crop value, all major grain crops, cotton, peanuts, and sugar occupied 17.7 million acres and had a value of $1.9 billion in 1964. At the same time, all horticultural crops, in-

TABLE 10.3.   *Value of cattle and dairy product sales in selected southern states, 1964*

State	Cattle and calves sold	Milk and cream sold
	(mil $)	
South Carolina	$ 19.7	$ 25.6
Georgia	73.7	50.1
Florida	71.5	86.1
Alabama	61.6	37.9
Mississippi	65.5	46.2
Southern Arkansas	12.8	1.9
Louisiana	42.2	47.3
East Texas	90.7	39.4
Total	$437.7	$334.5

Source: USDC, 1966 (4).

cluding Irish and sweet potatoes, had a total value of $653.2 million.

Another measure of the economic impact of the forage industry is sales of animal products directly resulting from forage utilization. Table 10.3 shows the sale value of cattle, calves, and dairy products by states in 1964 (4). The total of $772 million that resulted from the sales of $437.7 million of cattle and calves and $334.5 million of dairy products is considerably below the $950.5 million crop value placed on forages. Some of the difference between these two figures can be assigned to the value of forages that supported horses, sheep, and hogs. The acreage and value of forages devoted to these uses are not readily available.

It has been reported that on a nationwide basis dairy cattle obtain two-thirds and beef cattle three-fourths of their feed units from forages. This indicates that the true value of forage inputs to the ruminant livestock industry must be considerably less than the $772 million shown for sales of animals and animal products.

It seems most likely that much of the discrepancy between the apparent cash value of pasturage and harvested forages and the value of livestock sales lies in the inefficient conversion of grassland products to meat and milk. Even stored feeds like corn silage and high-quality hay are not completely recovered from storage and not completely consumed by animals. Poor quality hay is subject to much waste. Under the best grazing management practices, only 60–70 percent of the pasturage available to animals is eaten. In other situations where pastures are stocked with the number of cattle that can be supported during the driest season, great quantities of forage are wasted during the early spring period of lush growth.

Pasturage and stored forage cannot be sold for direct consumption by man, are not put under government loan, cannot be exported, and are of true economic value only when converted to milk or meat by ruminant animals or used for the support of other animals that bring profit or pleasure to man. Thus, all forage must be utilized by animals, and most of this will be in the immediate area where it is grown.

COSTS AND RETURNS FOR FORAGE
AND ROW CROP PRODUCTION

WHEN the individual farmer considers the establishment of a forage crop enterprise, he must compare the expected costs and returns with those of alternative crops he might grow. Data in Table 10.4 show the comparative costs of production and harvest and the total value per acre of a number of row crops and forage crops commonly grown in the southeastern part of the region. Several high-value crops not shown (including tobacco, peanuts, rice, and sugarcane) are likely to be more profitable than forages in those areas where they are adapted. However, acreage is controlled by government allotment or environmental requirements. Farmers do not have the option of growing these crops on all available acreage. Cotton is the only allotment-controlled crop listed. It is widely grown throughout much of the region except for Florida and the Gulf Coast, and many farmers either have an allotment or the opportunity of renting one if they have land suitable for cotton.

Data in Table 10.4 were obtained from a number of different sources and should be considered as average costs and returns for the region rather than specific budgets applicable to a given location. The data have been updated to reflect current nitrogen costs and commodity prices. The yield levels selected are considerably above the regional average but should be attained by above-average operators on land suitable for the crops being grown.

At the yield level shown, 670 pounds of lint per acre of cotton is the most profitable crop listed. This level is considerably higher than usually attained in the coastal plain areas, but it is often reached on the heavier soils in the northern part of the region and is easily surpassed on Mississippi delta soils.

With good management and adequate fertilization, perennial forage crops are more profitable than corn, soybeans, and small grains. On good land annual grazing crops are less profitable than corn or soybeans on a one-crop basis, but they are usually grown as part of a

TABLE 10.4. *Costs and returns for selected row and forage crops*

Crop	Yield/acre	Price/unit	Crop value	Preharvest cost	Harvest cost	Total cost	Return to land, management, and overhead
					($/acre)		
Cotton	670 lb lint	.388 lb	$219.70	$66.56	$73.76	$140.33	$103.52
	956 lb seed	.025 lb	24.14				
Corn	75 bu	1.15 bu	86.25	29.58	15.00	44.58	41.67
Soybeans	32 bu	2.25 bu	72.00	27.43	9.88	37.31	34.69
Barley	55 bu	1.00 bu	55.00	31.90	11.95	43.85	11.15
Corn silage	15 ton	10.00 ton	150.00	29.58	30.00	59.58	90.42
Coastal bermudagrass hay	6 ton	25.00 ton	150.00	29.51	45.00	74.51	75.49
Alfalfa hay	4.5	30.00 ton	135.00	38.47	33.75	72.22	62.78
Dallisgrass and white clover	*	‡	75.00	12.83	11.25	24.08	50.92
Bahiagrass and lespedeza	*	‡	75.00	13.76	11.25	25.01	49.99
Oats for grazing and hay	†	‡	75.00	30.66	11.25	41.91	33.09

* Estimated to provide 120 animal-days grazing with 1.25 lb gain per day and 1.5 ton hay/acre.
† Estimated to provide 120 animal-days grazing with 1.50 lb gain per day and 1.5 ton hay/acre.
‡ Dallisgrass and bahiagrass hay at $25/ton, oat hay at $20/ton, and beef gain at $0.25/lb.

double-cropping scheme where total annual per acre returns can be very favorable. On poor land the annual grazing crops are in a more favorable position when compared to corn or soybeans.

Corn or sorghum silages and Coastal bermudagrass hay are the most valuable forages because of their high per acre yields. At comparable management levels, row crop silage returns more profit than hay because harvesting is more highly mechanized and therefore cheaper.

Unless hay or silage is grown for sale, it is necessary to have an efficiently managed ruminant livestock enterprise for forages to return the profits these data indicate are possible. Pastures have sale value only as they are converted to meat or milk. Proper stocking rates, good animal health, efficiently managed breeding programs, and attention to market conditions are all necessary. If these factors are not properly controlled, even the best pastures will not be profitable.

FORAGE AND LIVESTOCK INDUSTRY TRENDS

THE PERIOD since 1945 has been one of almost revolutionary change for agriculture. Much of the mechanization and almost all the chemical technology we rely upon were developed during this period. Changes affecting the forage and livestock industry have been significant but less dramatic than changes in row crop production. See Table 10.5 (4).

Among developments that are increasing the efficiency of southern beef and milk production are the following:

1. The spread of Coastal bermudagrass and Pensacola bahiagrass across the humid South.
2. Introduction of the sorghum-sudangrass hybrids and cultivars of other annual and perennial forages.
3. Greater use of winter annual pastures (rye, ryegrass, clovers) for steer-fattening and dairy herds.
4. The growth of silage production and use for feeding dairy cattle, wintering beef breeding herds, and finishing slaughter cattle.
5. The drop in price of commercial fertilizer, particularly nitrogen.
6. The reduction in freight rates for midwestern feed grains.
7. The development of planned crossbreeding programs to utilize hybrid vigor in breeding programs.

TABLE 10.5. Trends in farmland usage in selected southern states, 1945–64

State*	Cropland harvested					Land pastured or harvested for hay				
	1945	1950	1954	1959	1964	1945	1950	1954	1959	1964
					(000 acres)					
South Carolina	4,149	3,960	3,393	2,691	2,263	2,461	2,718	3,131	2,281	2,104
Georgia	7,824	7,098	6,117	4,918	3,951	6,459	7,294	8,255	6,095	5,819
Florida	1,809	1,728	1,936	1,882	2,204	9,135	11,694	12,872	10,237	10,141
Alabama	6,163	5,729	4,812	3,715	2,990	6,660	7,479	9,211	7,304	7,161
Mississippi	6,473	6,136	5,530	4,564	4,400	8,508	9,024	10,745	9,482	9,472
Louisiana	3,490	3,149	3,011	2,426	2,673	4,211	5,033	6,106	5,834	5,895
Total	29,908	27,800	24,799	20,199	18,481	37,431	43,242	50,320	41,233	40,592

Source: USDC, 1964 (4).
* Data for years 1945–54 not readily available for southern Arkansas and eastern Texas because of partial state treatment.

During this same period the number of cattle and calves sold increased by 245 percent. The greatest part of this increase was represented by calf sales, and presumably many of these left the region for finishing in other areas. All states of the region shared in this increase; however, it was greatest in Louisiana where 373 percent more cattle and calves were sold in 1964 than in 1945. The cash value of all cattle sales also increased tremendously during this same period; however, long-term gross sales figures are not as meaningful as cattle numbers because of the wide fluctuations in market price. See Table 10.6 (4).

PROSPECTS FOR INCREASED USE OF FORAGES

A TASK FORCE studying research needs for forage, range, and pasture has estimated that meeting consumer demands for beef during the remainder of this century will require an additional 21 million head of cattle on pasture by 1980 and a doubling of present cattle populations by the year 2000 (2).

A look at the present level of fertilizer use on forage crops (Table 10.7) can give one indication of the potential for producing part of these additional millions of cattle on southern grasslands (4). Fertilizer use on forage is generally more common and the application rates are higher in the eastern part of the region. As a regional average, 27.6 percent of the nonforested pasture was fertilized in 1964. It can probably be assumed that little or no fertilizer was applied to woodland pastures. It is interesting to note, however, that the fertilized grasslands received almost as much fertilizer per acre as row crop land in the same states. No information is available on fertilizer analyses, so actual plant nutrient application rates cannot be accurately estimated.

In contrast to the fertilized acreages, a total of 8.7 million acres of hay and cropland pasture and 19.1 million acres of other nonforested grassland received no fertilizer at all in 1964. When proper species are present, forage yields can easily be doubled with moderate fertilization; it is usually economical to fertilize for even greater increases if the increased production can be utilized.

It is quite likely that some of the more open areas within the 20.6 million acres of woodland pasture could be made to support greater numbers of cattle by moderate fertilization. On many of

TABLE 10.6. *Trends in cattle numbers and sales in selected southern states, 1945–64*

State*	Cattle and calves on farms					Cattle and calves sold				
	1945	1950	1954	1959	1964	1945	1950	1954	1959	1964
					(000 head)					
South Carolina	389	370	615	497	555	88	109	204	214	221
Georgia	1,140	1,003	1,626	1,353	1,747	280	314	574	602	765
Florida	1,115	1,101	1,647	1,501	1,822	216	368	659	664	807
Alabama	1,282	1,269	1,796	1,526	1,848	415	439	654	635	727
Mississippi	1,655	1,569	2,320	2,006	2,351	370	467	751	784	864
Louisiana	1,475	1,285	1,850	1,656	1,867	276	366	590	616	647
Total	7,056	6,597	9,854	8,539	10,190	1,645	2,063	3,432	3,515	4,031

Source: USDC, 1966 (4).
* Data for years 1945–54 not readily available for southern Arkansas and eastern Texas because of partial state treatment.

TABLE 10.7. *Summary of fertilizer usage in selected southern states, 1964*

State	Hay and cropland pasture		Other pasture*		Other major field crop†	
	Acreage fertilized	Rate per acre	Acreage fertilized	Rate per acre	Acreage fertilized	Rate per acre
	(% of total)	*(lb)*	*(% of total)*	*(lb)*	*(% of total)*	*(lb)*
South						
Carolina	54.2	569	22.7	471	85.8	598
Georgia	45.0	556	12.7	463	94.1	604
Florida	35.2	452	22.7	317	91.9	457
Alabama	33.3	468	18.2	389	94.7	497
Mississippi	24.3	394	10.9	328	95.1	385
Southern						
Arkansas	15.6	270	7.9	275	25.1	175
Louisiana	16.8	336	9.0	340	96.1	287
East Texas	19.9	308	10.0	199	76.1	227
Regional						
average	27.6	437	13.9	336	85.9	453

Source: USDC, 1966 (4).
* Does not include woodland pasture.
† Includes two or three principal crops such as corn, cotton, soybeans, sorghum, or rice.

these wooded areas, part of the fertilizer cost could be recovered by increased tree growth, thus offering greater chance for profit.

There is no way to estimate the amount of grassland that requires renovation or introduction of new species, but it is no doubt significant. For instance, approximately 2 million acres of Coastal bermudagrass are reported in the region excluding Texas, Arkansas, and Florida. This same area has some 2.4 million acres of perennial hayland and 16.4 million acres of open pasture. Because Coastal bermudagrass is high yielding, drought resistant, responsive to fertilizer, and produces good quality hay, it seems reasonable to believe that most of the hay acreage and at least one-third of the pasture acreage of this region should be converted to Coastal. This would require an increase of some 5.5 million acres of this high-yielding forage. A significant part of the remaining acreage that should be growing cool-season species probably needs to be reseeded to high-yielding grass-legume mixtures or, in the case of much of the existing tall fescue sod, should be overseeded with white, ladino, or red clover.

The possibilities for increasing per acre forage yields by renovation, reestablishment with higher yielding species, overseeding bermudagrass or bahiagrass with winter annual clovers, or proper liming and fertilization indicate that southern grasslands should support at least twice as many cattle without any new technological developments.

A number of animal scientists and economists in the humid

South were asked to predict future changes in the cattle industry. All predicted continued expansion in beef production. None predicted revolutionary changes in production methods but rather a general improvement in management and gradual increases in the use of corn and sorghum silage for both wintering and finishing cattle and in the feeding of grain on pasture.

Scientists from South Carolina and Georgia predicted the greatest increase in silage use, while the strongest interest in supplementing pasture with grain was reported from Louisiana. Florida workers predicted moderate increases in use of both silage and grain with grass feeding. Predictions for eastern Texas included continued reliance on permanent pasture, with conversion of much existing row crop acreage to pasture. The interest in winter annual grazing crops for young cattle is greatest in the southern part of the region. No one predicted that finishing cattle in the feedlot on high grain rations would become a major enterprise in the region; however, there is localized interest in several states in finishing cattle on all-silage rations.

It seems that a shortage of capital is one of the factors that most limits a rapid buildup of the southern cattle industry and, consequently, the contribution of forages. A rapid infusion of outside capital into the feedlot and packing plant industry, such as has already happened in the Southwest, could have a drastic effect on the economics of feeder calf production. This would provide a ready market for home-grown calves without transportation costs for shipping out of the area as feeders and back in as finished meat. In addition, such feedlots could provide a tremendous market for silage grown on a contract basis.

SUMMARY

In summary, the forage crop—ruminant livestock industry in the humid South utilizes considerably more than half the nonforested farmland, results in greater total sales volume than all horticultural crops, and produces 40 percent as much in total sales as all row crops and close-grown field crops. The livestock industry of the region seems destined for greater future expansion than row crop agriculture because of greater product demand and the tremendous opportunities for increasing production per acre.

REFERENCES

1. Klingebiel, A. A., and P. H. Montgomery. 1961. Land-capability classification. SCS, USDA Agr. Handbook 210.

2. USDA. 1967. A national program of research for forage, range, and pasture. Report of joint task force of the USDA, the state universities, and land-grant colleges.

3. USDA, Soil Conservation Service. 1967. Conservation needs inventory.

4. USDC. 1966. 1964 U.S. census of agriculture. Bureau of Census.

11

SIGNIFICANCE OF FORAGES TO THE ECONOMY OF THE NORTH CENTRAL REGION

MAURICE E. HEATH

FORAGES as referred to here include all pasture, hay, silage, green chop, and crop residues or roughages that are consumed by farm animals. The market for forages is the farmer's livestock. How the livestock farmer manages his soil-forage-animal resources determines the extent and profitability of his forage market (1).

In the north central region (including the states of Illinois, Indiana, Iowa, Michigan, Minnesota, Missouri, Ohio, and Wisconsin) forages contribute 47.8 percent of all the feed units consumed by the livestock industry as a whole. A feed unit is equivalent in feeding value to a pound (or ton) of corn (2). Pasture represents 27.2 percent, harvested forage 20.6 percent, and concentrates 52.2 percent of all livestock feed units used in the region (Fig. 11.1). Gross agricultural income from livestock sales is nearly twice that from crop sales.

CROPS AND CROP SHIFTS

THE NORTH CENTRAL REGION represents 15 percent of the land area of the 48 states. In acreage, the area produces 63.7 percent of the corn, 24.8 percent of the small grain, 66.2 percent of the soybeans, 31.5 percent of the hay, and 8 percent of the pasture of the United States (Table 11.1) (6). The only record of pasture acreage is the U.S. census, which is compiled every five years. The 1964 U.S. census has been used for base data. Crop, livestock, and income trends have been calculated from USDA agricultural statistics. For example, in 1964 the region produced 20.6 million acres of harvested hay. By 1970

140

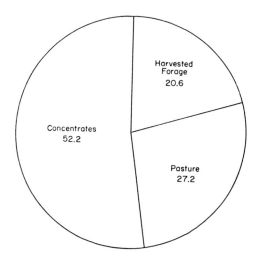

FIG. 11.1. PERCENTAGE OF ALL FEED UNITS OF THE NORTH CENTRAL REGION CONSUMED BY ALL LIVESTOCK AS CONCENTRATES, HARVESTED FORAGE, AND PASTURE. (SEE TABLE 11.3.)

there was a decrease of 13.3 percent in acreage of harvested hay or a total of 17.8 million acres.

The largest acreage shift for any one crop has been in soybeans, a 1970 acreage increase of 28.2 percent over 1964. The 5.6 million acre increase in soybeans is about equal to the combined hay and small-grain acreage reduction. Further reductions in pasture acreage (3.1 million acres less in 1964 compared to 1959) account for the total acreage shifts to row crops.

The primary Corn Belt area (Fig. 11.2) dominates the central east-west area of the region (5). In this same area the greater part of the soybean acreage is also grown and accelerated shifts to more row crops have been observed.

SILAGE CROPS

In 1970, corn harvested as silage amounted to 3.4 million acres; this was 7.9 percent of the total corn acreage of the region compared to 9.8 percent in 1964. Wisconsin, leading the region, harvested 934,000 acres of corn for silage, which was a third of all the corn grown in the state. Minnesota was second in corn silage production with 676,000 acres, and Iowa was third with 583,000 acres. There were 46,000

TABLE 11.1. *Acreage in north central region devoted to forages and field crops, 1964*

						Pasturelands			
State	Corn	Sorghum	Small grains	Soybeans	All hay*	Cropland only, pastured	Woodland, pastured	Other pasture	Total
					(000 acres)				
Illinois	9,461	22	3,019	5,600	1,806	1,669	1,501	1,746	4,916
Indiana	4,480	15	1,794	2,680	1,210	1,335	1,018	1,102	3,455
Iowa	10,127	55	2,346	4,177	3,201	2,632	1,551	3,248	7,431
Michigan	1,880	1	1,659	317	1,828	1,147	908	577	2,632
Minnesota	5,642		6,006	2,788	3,540	1,307	2,410	2,118	5,835
Missouri	3,135	287	1,837	2,627	2,920	4,428	5,214	5,835	15,477
Ohio	3,246	2	2,078	1,761	1,945	1,164	930	2,177	4,271
Wisconsin	2,464	…	2,032	135	4,131	1,765	2,802	1,845	6,412
Regional total	40,435	383	20,771	20,085	20,581	15,447	16,334	18,648	50,429
% of 1964 U.S. total	63.7	2.6	24.8	66.2	31.5	26.9	19.9	3.8	8.0
1970 % increase over 1964†	5.2	2.3	−13.7	28.2	−13.3	…	…	…	…

Source: USDC, 1966 (6).
* Includes acres for grass silage.
† USDA agricultural statistics.

FIG. 11.2. THE CORN BELT, OR PRIMARY ADAPTATION OF CORN,
IN RELATION TO THE STATES OF THE NORTH CENTRAL REGION.

acres of sorghum for silage reported in 1970. Over half the sorghum
silage acreage was grown in Missouri.

Corn is probably the most popular crop for silage. In general,
it is difficult to make poor corn silage when compared to some of the
problems of making grass silage. However, 882,000 acres of grass
were harvested for silage in 1964, which represents 48.8 percent of all
grass silage made in the United States. The acres of grass harvested
for silage was 4.3 percent of all the hay acreage reported for the region.

LEGUMES

IN 1970, 10.4 million acres of alfalfa and alfalfa-grass mixtures were
grown for hay in the region compared to 5.5 million acres of clover
and clover-grass mixtures. Alfalfa is the preferred hay legume on
well-drained soils and has replaced much of the red clover acreage
since the midforties. Wisconsin has nearly 3 million acres of alfalfa,
Minnesota over 2 million acres, and Michigan and Iowa each have
over a million acres. However, red clover–grass mixtures are still
popular on many farms for short rotations and for use in the livestock

program. Red clover is also important on many soils not well suited to the production of alfalfa.

Other important legumes include white clover, which volunteers in many of the permanent pastures throughout the region; ladino clover, which is often seeded in special purpose mixtures for use with swine and dairy cattle; birdsfoot trefoil, which has special adaptation in permanent hill land pastures through the Corn Belt and areas to the north; and Korean and common lespedeza, which as reseeding annuals fill an important midsummer pasture need in the southern part of the region.

GRASSES

OVER 34 million acres of permanent pasture are reported for the region. The predominant grass species on much of this grassland is Kentucky bluegrass. However, since 1945 tall fescue has become a major grass over much of the southern area. Timothy is commonly used with red clover for hay and pasture. Smooth bromegrass, orchardgrass, and timothy are frequently used with alfalfa. Reed canarygrass has been used in wet areas, and pasture trials indicate promise for use on upland soils where proper fertility and management are provided.

CROP RESIDUES

CORNSTALKS and small-grain fields following grain harvest have always been sources of cheap maintenance feed for ruminant livestock in the Corn Belt. In 1970 the region produced 39 million acres of cornstalks with a potential dry-matter yield of 2–3 tons per acre. Also available were 17 million acres of small-grain stubble fields, many of which contained new grass-legume seedings. Many farmers look to this source of low-cost feed in fall and winter for their beef cow herds.

Where corn farms are becoming larger and more specialized in grain production, there appears to be less consideration and time for utilization of crop residues by cattle.

LAND USE

PERENNIAL GRASSES and legumes provide a crop on land not suitable for intensive grain production. On many farms this land provides

TABLE 11.2. *Livestock on farms in the states of the north central region, 1964*

State	All cattle	Milk cows	All hogs	Sheep
		(000 head)		
Illinois	3,751	422	7,309	561
Indiana	2,053	328	4,497	361
Iowa	7,285	736	13,674	1,365
Michigan	1,725	574	716	323
Minnesota	4,281	1,237	3,396	774
Missouri	4,522	461	3,854	447
Ohio	2,164	544	2,513	838
Wisconsin	4,384	2,083	1,735	203
Regional total	30,165	6,385	37,694	4,872
% of 1964 U.S. total	28.6	43.7	69.7	19.1
1970 % increase over 1964*	—2.3	—20.0	15.0	—33.6

Source: USDC, 1966 (6).
* USDA agricultural statistics.

pasture for beef cattle in summer. After grain harvest the cattle are allowed to glean the stubble and cornstalk fields. Such management practices are complementary and provide a more intensive use of total resources. New forage management and cultural practices are contributing to greater intensification in what has been considered one of the more difficult soil-climate areas.

LIVESTOCK AND SOURCE OF FEED

In 1964 the north central region reported 28.6 percent of all the cattle in the United States, 69.7 percent of the hogs, and 19.1 percent of the sheep (Table 11.2) (6). However, between 1964 and 1970 we observe a 2.3 percent decrease in cattle numbers, a 15.0 percent increase in hogs, and a 33.6 percent decrease in sheep for the region. One of the big shifts has been from dairy to beef cattle. However, the increase in beef cattle was not sufficient to overcome the slight decrease in total cattle numbers.

Iowa is the leading state in the region in numbers of cattle, hogs, and sheep. Wisconsin leads in numbers of dairy cattle. The tons of feed units to support the regional livestock industry as a whole are reflected in Table 11.3 (3). Note that 41.4 percent of all concentrate feed units in the United States is consumed by the livestock industry of the north central region, 37.5 percent of all harvested forage, and 25.3 percent of the feed units of all pasture.

Iowa and Wisconsin are the two largest forage-using states in the region, followed by Missouri and Minnesota. For the region as a whole, 47.8 percent of all feed units consumed by the livestock indus-

TABLE 11.3. *Feed units consumed by all livestock for the states of the north central region as concentrates, harvested forage, and pasture, 1966–67*

State	Concentrates	Harvested forage	Pasture	Total	Concentrates	Harvested forage	Pasture	Total
	(000 tons)				(%)			
Illinois	11,282	2,571	4,607	18,460	61.1	13.9	25.0	100.0
Indiana	6,712	1,670	3,071	11,453	58.6	14.6	26.8	100.0
Iowa	21,638	4,633	10,145	36,416	59.4	12.7	27.9	100.0
Michigan	2,898	2,114	1,494	6,506	44.5	32.5	23.0	100.0
Minnesota	9,630	5,200	3,874	18,704	51.5	27.8	20.7	100.0
Missouri	7,135	2,818	6,934	16,887	42.2	16.7	41.1	100.0
Ohio	5,098	2,181	2,868	10,147	50.2	21.5	28.3	100.0
Wisconsin	7,143	7,127	4,339	18,609	38.4	38.3	23.3	100.0
Regional total	71,536	28,314	37,332	137,182	52.2	20.6	27.2	100.0
% U.S. total	41.4	37.5	25.3	34.6				

Source: Hodges, 1971 (3).

146

TABLE 11.4. *Feed units consumed by various livestock classes by states of the north central region as concentrates, harvested forage, and pasture, 1966–67*

Livestock class	Concentrates	Harvested forage	Pasture	Total	Concentrates	Harvested forage	Pasture	Total
	(000 tons feed units)				(%)			
Milk cows	11,321	13,543	5,466	30,330	37.3	44.7	18.0	100.0
Other dairy cattle	956	2,992	1,930	5,878	16.3	50.9	32.8	100.0
Cattle on feed	12,643	4,166	…	16,809	75.2	24.8	…	100.0
Other beef cattle	2,098	6,904	20,453	29,455	7.1	23.4	69.5	100.0
Hens and pullets	4,339	…	…	4,339	100.0	…	…	100.0
Chickens raised	975	…	…	975	100.0	…	…	100.0
Broilers	464	…	…	464	100.0	…	…	100.0
Turkeys	2,474	…	…	204	100.0	…	…	100.0
Hogs	35,702	…	7,516	43,218	82.6	…	17.4	100.0
Horses and mules	260	538	637	1,435	18.1	37.5	44.4	100.0
Stock sheep	255	132	1,330	1,717	14.9	7.7	77.4	100.0
Sheep on feed	49	39	…	88	55.7	44.3	…	100.0
Regional total	71,536	28,314	37,332	137,182	52.2	20.6	27.2	100.0

Source: Hodges, 1971 (2).

147

try is derived from forages. Harvested forages furnish 20.6 percent of the feed units, and pastures furnish 27.2 percent.

FORAGES USED BY LIVESTOCK CLASSES

As SHOWN in Table 11.4 (3), the beef cow-calf and stocker (other beef cattle) enterprise of the region not only represents the greatest tonnage of forage feed units consumed but also the largest amount of forage used in relation to concentrates; harvested forage and pasture represent 92.9 percent of total feed units consumed. Milk cows are second with 62.7 percent of their feed units from forage, and hogs are third with 17.4 percent. Sheep, horses, and mules provide the least market for forage.

Figure 11.3 illustrates more graphically the importance of each livestock class in the region as related to the total forage market. For example, dairy cows provide a market for 28.95 percent of all forage feed units consumed in the region; beef cattle, other than cattle on feed, provide 41.7 percent of the market.

FORAGE VALUE AND LIVESTOCK SALES

THE ANNUAL gross livestock sales are about twice the crop sales for the region. In 1970 the gross livestock sales for the region totaled $10.1

TABLE 11.5. *The total forage feed units consumed by livestock, value of the forage, 1964 gross livestock sales, and percent of feed units from forage by states for the north central region*

State	Forage feed units*	Estimated forage value†	Livestock sales	Livestock feed units from forage*
	(000 tons)	*(mil $)*		*(%)*
Illinois	7,178	$ 287.1	$1,215.1	38.9
Indiana	4,741	189.6	739.7	41.4
Iowa	14,778	591.1	2,368.8	40.6
Michigan	3,608	144.3	437.0	55.5
Minnesota	9,074	363.0	1,132.7	48.5
Missouri	9,752	390.1	839.1	57.8
Ohio	5,049	202.0	671.8	49.8
Wisconsin	11,466	458.6	1,075.4	51.6
Regional total	65,646	$2,625.8	$8,479.6	47.8
% of U.S. total	29.4	. . .	38.6	. . .
1970 % regional increase over 1964	. . .	. . .	19.0	. . .

* From Table 11.3.
† Calculated on the basis of $40 per ton for corn. Forage feed unit tons × $40.

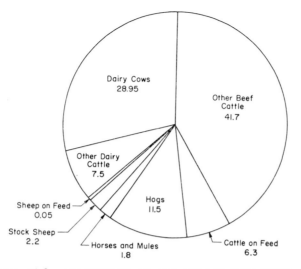

FIG. 11.3. PERCENTAGE OF ALL FORAGE FEED UNITS (PASTURE
PLUS HARVESTED FORAGE) CONSUMED BY EACH LIVESTOCK CLASS
IN THE NORTH CENTRAL REGION. (SEE TABLE 11.4.)

billion. To support this industry, 47.8 percent of the feed units were
from forages and 52.2 percent from concentrates. The forage portion
of the total feed units used in each of the eight states varied from a
low of 38.9 percent for Illinois to a high of 57.8 percent for Missouri.

Three methods of placing a dollar value on forage (feed unit
value, value for hay equivalents, and percent of livestock sales as related to forage feed units consumed) were compared by Hodgson (4).
He concluded that the method of equating the value of a forage feed
unit to the value of a corn feed unit was the most accurate. Using
this method (with corn valued at $40 per ton [$1.12/bu 1966–67]) we
find the total value of the 65.6 million tons of forage feed units utilized
in the region to be $2.6 billion (Table 11.5). The dollar value for
forage, when compared to the market price of corn, represents 30.9
percent of the gross livestock sales for the region.

SUMMARY

FORAGES are a major source of feed for the livestock industry of the
north central region, contributing 47.8 percent of the total feed units
consumed. Beef cattle represent the largest market for forages, followed by dairy cattle. On many farms, forages are complementary to

the grain–livestock–land use production programs and contribute to the intensification of total resources. Forages have major economic value because they are utilized by livestock and marketed through livestock. When the value of the utilized forage feed units are equated with the market price of corn feed units, this amounts to 30.9 percent of the 1964 livestock sales, or $2.6 billion for the region.

REFERENCES

1. Heath, M. E. 1973. Hay and pasture seedings for the Central and Lake States. *In* Forages. Iowa State Univ. Press, Ames.

2. Hodges, E. F. 1963. Livestock-feed relationships 1909–1963. USDA Stat. Bull. 337.

3. ———. 1971. Preliminary feed unit data for the 1966–67 period. USDA-ERS. Correspondence.

4. Hodgson, H. J. 1968. The importance of forages in livestock production in the United States. Am. Soc. Agron. Spec. Publ. 13.

5. USDA and North Central Agr. Exp. Stas. cooperating. 1960. Major soils of the North Central region. Wis. Agr. Exp. Sta. Bull. 544.

6. USDC. 1966. 1964 U.S. census of agriculture. Bureau of Census.

12

IMPACT OF FORAGES ON THE AGRICULTURAL INCOME OF THE GREAT PLAINS

DONALD   F.   BURZLAFF

THE HETEROGENEITY of climate, soils, and relief of the Great Plains creates a unique blend of livestock and crop enterprises within the region. The balance in forage, concentrate, and livestock production is reflected in the agricultural income of the Great Plains states. The land area represents 31 percent of the total of the 48 conterminous states of the United States (Fig. 12.1). Great Plains farms and ranches represent approximately 50 percent of the total agricultural lands of the United States. In 1968, 23.7 percent of the total U.S. agricultural income from crop and livestock marketing was attributed to the ten states of the Great Plains.

LIVESTOCK AND FORAGE RESOURCES

IN 1968 Great Plains states supported 39 percent of the total cattle; 49.4 percent of the beef cows, two years of age or older; and 57 percent of the stock sheep and lambs (Table 12.1) (2, 3). The increase in cattle numbers and in beef cows has been common to all states in the region. States with the greatest increases are those where feeding enterprises have increased greatly. The total number of stock sheep have declined, and the feed resources have been diverted to the increased cattle numbers. The increase in beef cow numbers also reflects the decrease in the amount of dairy cows during the past decade.

The forages fed to cattle, sheep, and other livestock in the Plains are produced by some 461.7 million acres of grazing lands and 26.5 million acres of harvested forages (Table 12.2) (3). Approximately 82 percent of the grazing land is privately owned. The remainder is

FIG. 12.1.   THE GREAT PLAINS STATES.

TABLE 12.1.   *Livestock populations of the Great Plains and their relationships to those of the 48 United States: A comparison of data from 1958 and 1968*

State	Total cattle number		Beef cows 2 years or older		Stock sheep	
	1958	1968	1958	1968	1958	1968
	(000 head)					
North Dakota	1,870	2,132	600	956	559	329
South Dakota	3,230	4,323	1,205	1,638	1,249	1,096
Nebraska	4,630	6,394	1,395	1,907	308	241
Kansas	3,961	5,564	1,062	1,676	479	333
Oklahoma	2,958	4,480	1,192	2,000	207	114
Texas	7,736	10,972	3,604	5,356	4,700	3,986
Montana	2,247	2,984	1,105	1,556	1,604	1,165
Wyoming	1,140	1,447	511	677	2,059	1,697
Colorado	1,979	3,060	704	995	1,252	884
New Mexico	1,056	1,346	581	690	1,172	823
Total	30,807	42,602	11,959	17,451	13,589	10,668
U.S. total	93,350	108,902	24,287	35,314	27,327	19,078
% of U.S. total	33	39.1	49.2	49.4	49.7	56.9

Sources: USDA, 1959, 1969 (2, 3).

TABLE 12.2. *The forage resources of the Great Plains, 1968*

State	Harvested forages	Grazing lands Private	Public
		(000 acres)	
North Dakota	3,727	14,073	. . .
South Dakota	5,206	26,573	524
Nebraska	4,797	25,119	160
Kansas	3,044	20,340	345
Oklahoma	1,642	23,813	3,680
Texas	2,566	111,609	14,594
Montana	2,236	50,278	10,802
Wyoming	1,209	34,085	15,698
Colorado	1,726	27,619	15,142
New Mexico	310	44,806	22,533
Total	26,463	378,314	83,478

Source: USDA, 1969 (3).

public land (under the management of various state and federal agencies) that is leased for grazing purposes.

AGRICULTURAL INCOME

THE IMPACT of these livestock numbers is evident in the total agricultural income of the area. The income from livestock and crops has risen 22 percent from 1958 to 1968 (Tables 12.3 and 12.4) (2, 3). Cash sales from crops are lower. This reflects the increased activity in feeding and the marketing, through livestock, of crops produced

TABLE 12.3. *Cash sales of livestock and crops as they contribute to total agricultural income of the Great Plains, 1958*

State	Cash sales livestock	Cash sales crops	Total agricultural income
		(000 $)	
North Dakota	$ 219,354	$ 438,944	$ 658,298
South Dakota	478,912	206,011	684,923
Nebraska	781,108	466,636	1,247,744
Kansas	563,662	619,936	1,183,598
Oklahoma	321,597	312,180	633,777
Texas	929,587	1,443,033	2,372,620
Montana	242,277	198,647	440,924
Wyoming	135,461	32,483	167,944
Colorado	341,151	227,354	568,505
New Mexico	141,295	83,990	225,285
Total	$4,154,404	$4,029,214	$8,183,618

Source: USDA, 1959, Table 690 (2).

TABLE 12.4.   *Cash sales of livestock and crops as they contribute to total agricultural income of the Great Plains, 1968*

State	Cash sales livestock	Cash sales crops	Total agricultural income
		(000 $)	
North Dakota	$ 266,998	$ 453,316	$ 720,314
South Dakota	753,001	205,034	958,035
Nebraska	1,273,329	484,227	1,757,556
Kansas	1,002,036	533,878	1,535,914
Oklahoma	578,305	267,678	845,903
Texas	1,414,115	1,254,916	2,669,031
Montana	316,184	193,321	509,505
Wyoming	204,608	37,439	242,047
Colorado	686,537	217,058	903,595
New Mexico	220,773	101,580	322,353
Total	$6,715,886	$3,748,447	$10,454,253

Source: USDA, 1969, Table 679 (3).

in the Plains. The cash income from 1968 livestock sales was approximately 1.79 times as large as cash sales from crops. These cash livestock sales represent in part a return from concentrates and forages marketed through livestock consumption.

The proportion that each kind of livestock contributed to total agricultural income was calculated for 1958 and 1968. These values are shown in Tables 12.5 (2) and 12.6 (3). Sale of beef cattle made up

TABLE 12.5.   *Contribution of various kinds of livestock to adjusted cash sales of livestock in the Great Plains, 1958*

State	Beef cattle*†	Sheep*	Other livestock*	Dairy products	Total cash sales*
			(000 $)		
North Dakota	$ 127,773	$ 7,403	$ 42,022	$ 42,000	$ 214,198
South Dakota	229,581	12,138	199,166	36,997	477,882
Nebraska	324,222	7,667	388,010	60,643	780,542
Kansas	192,283	5,940	290,761	73,308	562,292
Oklahoma	166,819	2,174	89,296	63,308	321,597
Texas	398,239	19,854	344,261	167,110	929,464
Montana	176,354	13,946	33,000	18,850	242,150
Wyoming	87,927	17,268	22,468	7,684	135,347
Colorado	137,040	17,146	150,227	36,720	341,133
New Mexico	68,851	6,111	52,637	13,143	140,742
Total	$1,909,089	$109,647	$1,611,848	$519,763	$4,145,347

Source: USDA, 1959 (2).
  * Adjusted by subtracting the purchase cost of cattle, sheep, and hogs brought into each particular state.
  † Includes dairy cattle slaughtered as beef.

TABLE 12.6. *Contribution of various kinds of livestock to adjusted cash sales of livestock in the Great Plains, 1968*

State	Beef cattle*†	Sheep*	Other livestock*	Dairy products	Total cash sales*
			(000 $)		
North Dakota	$ 183,194	$ 5,634	$ 27,085	$ 41,438	$ 257,351
South Dakota	397,261	18,093	174,261	67,404	657,019
Nebraska	639,680	7,679	246,790	69,678	963,827
Kansas	499,830	5,912	120,274	88,426	714,442
Oklahoma	330,396	1,636	48,139	71,078	451,249
Texas	663,648	32,846	266,894	191,743	1,155,131
Montana	245,175	11,960	24,012	17,515	298,662
Wyoming	130,098	18,169	10,368	7,789	166,424
Colorado	336,823	20,143	31,128	51,231	439,325
New Mexico	117,031	5,972	12,152	21,074	156,229
Total	$3,543,136	$128,044	$961,103	$627,376	$5,259,659

Source: USDA, 1969 (3).
* Adjusted by subtracting the purchase cost of cattle, sheep, and hogs brought into each particular state.
† Includes dairy cattle slaughtered as beef.

67 percent of the adjusted cash sales of livestock products in 1968 compared to 46 percent in 1958.

Contribution of Forages. Of the total feed units consumed by livestock in 1967, 54.3 percent came from forages and 45.7 percent from concentrates. These values represent averages for the United States (2). The percentage for a specific state would be higher or lower depending upon the proportion that monogastric livestock and poultry contribute to the total income. It would be further modified by the extent of the livestock-feeding enterprise in the state.

For calculations presented in this chapter the percentage was varied according to the proportion that feeder cattle contributed to total livestock sales. In Nebraska, Kansas, Oklahoma, Texas, and Colorado, where feeding is a major livestock enterprise, the contribution of forages to total feed units consumed by beef cattle sold in 1968 was considered to be 55 percent. In the remaining five states of the Great Plains this was considered to be 70 percent because of the increased proportion of beef cattle sold directly from the range. The amount that forages contributed to the feed units consumed by sheep was considered to be 60 percent; by dairy cows, 33 percent; and other livestock (including horses, hogs, and poultry), 5 percent.

In 1958, calculations were based on the same percentages except that only Nebraska, Texas, and Colorado were considered to be in-

TABLE 12.7.   *The amount of adjusted cash sales of livestock attributed to forages in the Great Plains, 1958*

State	Beef cattle*†	Sheep‡	Other livestock§	Dairy‖	Total
			(000 $)		
North Dakota*	$ 62,609	$ 3,109	$ 1,260	$ 7,560	$ 74,538
South Dakota*	112,495	5,098	5,975	6,659	130,227
Nebraska†	126,447	3,220	11,640	10,915	152,222
Kansas*	94,219	2,495	8,723	13,195	118,632
Oklahoma*	81,741	913	2,679	11,395	96,728
Texas†	155,313	8,339	10,327	30,080	204,059
Montana*	86,413	5,857	990	3,393	96,653
Wyoming*	43,084	7,252	674	1,383	52,393
Colorado†	53,446	7,201	4,507	6,610	71,764
New Mexico*	33,736	2,567	1,579	2,366	40,248
Total	$849,503	$46,051	$48,354	$93,556	$1,037,464

* The estimate obtained by multiplying the adjusted cash sales for beef cattle by 70% (amount of adjusted sales that reflect feed costs). This value multiplied by 70%. This represents the proportion of feed units consumed that are provided by forage in these states.
† Adjusted cash sales × 70% × 55%.
‡ Adjusted cash sales × 70% × 60%.
§ Adjusted cash sales × 70% × .04½%.
‖ Adjusted cash sales × 50% × 33%.

volved in major feeding operations. Assuming that 70 percent of the cash sales value of cattle represents feed cost, the feed units consumed by livestock that are attributed to forages can be calculated by multiplying the total cash sales (adjusted) by this percentage, times the appropriate percentage mentioned above. The procedure for calculation of forage values follows that outlined by Hodgson (1). Approximately 70 percent of the cash sale value of sheep represents feed costs.

When the portions of adjusted cash sales representing forages have been calculated, it is evident that beef cattle are major forage consumers (Tables 12.7 and 12.8). They consume about 82–88 percent of the forage produced in the Great Plains. Forage values in the Great Plains represent about 25 percent of the value of total adjusted income from livestock sales in 1958. This increased to 33 percent in 1968, indicating forages were playing an increasingly important role in livestock production.

In 1958, harvested forages represented 52 percent of total forage value. Pasture and range accounted for the balance (Table 12.9). By 1968, harvested forages accounted for 58 percent of total forage value (Table 12.10), while pasture and range represented only 42 percent even though the total value of both had increased substantially.

TABLE 12.8.   *The amount of adjusted cash sales of livestock attributed to forages in the Great Plains, 1968*

State	Beef cattle*†	Sheep‡	Other livestock§	Dairy‖	Total
			(000 $)		
North Dakota	$ 89,765*	$ 2,366	$ 812	$ 7,666	$ 100,609
South Dakota	194,658*	7,599	5,227	12,470	219,954
Nebraska	249,475†	3,225	7,404	12,890	272,994
Kansas	194,934‡	2,483	3,608	16,358	217,383
Oklahoma	128,854†	687	1,444	13,149	144,134
Texas	258,823†	13,795	8,006	35,472	316,096
Montana	120,136*	5,023	720	3,240	129,119
Wyoming	63,748*	7,631	311	1,440	73,170
Colorado	131,361†	8,460	934	9,478	150,233
New Mexico	57,345*	2,508	365	3,899	64,117
Total	$1,492,099	$53,777	$28,831	$112,777	$1,687,809

* The estimate obtained by multiplying the adjusted cash sales for beef cattle by 70% (amount of adjusted sales that reflect feed costs). This value multiplied by 70%. This represents the proportion of feed units consumed that are provided by forage in these states.
† Adjusted cash sales $\times$ 70% $\times$ 55%.
‡ Adjusted cash sales $\times$ 70% $\times$ 60%.
§ Adjusted cash sales $\times$ 70% $\times$ .04½%.
‖ Adjusted cash sales $\times$ 50% $\times$ 33%.

TABLE 12.9.   *Value of forages, range, and pasture in the Great Plains, 1958*

State	Total forage value*	Harvested forage value†	Value of pasture and range‡
	(000 $)		
North Dakota	$ 74,538	$ 64,653	$ 9,885
South Dakota	130,227	74,980	55,247
Nebraska	152,222	78,651	73,571
Kansas	118,632	83,456	35,176
Oklahoma	96,728	36,100	60,628
Texas	204,059	56,332	147,727
Montana	96,653	53,020	43,633
Wyoming	52,393	24,992	27,401
Colorado	71,764	52,607	19,157
New Mexico	40,248	15,113	25,135
Total	$1,037,464	$539,904	$497,560

* From Table 12.7.
† USDA, 1959 (3).
‡ Column 1 minus column 2.

TABLE 12.10. *Value of forages, range, and pasture in the Great Plains, 1968*

State	Total forage value*	Harvested forage value†	Value of pasture and range‡
		(000 $)	
North Dakota	$ 100,609	$ 72,448	$ 28,161
South Dakota	219,954	140,673	79,281
Nebraska	272,994	170,522	102,472
Kansas	217,383	148,387	68,996
Oklahoma	144,134	68,158	75,976
Texas	316,096	124,958	191,138
Montana	129,119	85,012	44,107
Wyoming	73,170	36,735	36,445
Colorado	150,233	104,885	45,348
New Mexico	64,117	31,740	32,377
Total	$1,687,809	$983,518	$704,291

* From Table 12.8.
† USDA, 1969 (3).
‡ Column 1 minus column 2.

The total value of harvested forage in the Great Plains states approached $1 billion for 1968. See Tables 12.11 (2) and 12.12 (3). Hay of all classes represented 79 percent of this or $781.2 million, corn silage 16 percent or $156.6 million, and sorghum silage 5 percent or $45.6 million. This represents an increase in tons of harvested forage over the 1958–68 decade. The production of corn silage more than doubled in this time interval. This is another indication that Great Plains farmers are interested in harvesting the maximum total digestible nutrients produced by their crops.

TABLE 12.11. *Volume and value of harvested forages in the Great Plains, 1958*

State	Corn silage*		Sorghum silage†		Hay (all)	
	(000 tons)	*(000 $)*	*(000 tons)*	*(000 $)*	*(000 tons)*	*(000 $)*
North Dakota	2,324	$14,944	2	$ 10	3,823	$ 49,699
South Dakota	1,950	11,700	200	1,000	5,190	62,280
Nebraska	820	4,920	627	3,135	7,844	70,596
Kansas	1,179	7,074	4,685	23,425	4,605	52,957
Oklahoma	81	486	805	4,025	2,038	31,589
Texas	432	2,592	800	4,000	2,487	49,740
Montana	348	2,088	...	...	2,996	50,932
Wyoming	285	1,710	...	...	1,663	23,282
Colorado	1,833	10,998	175	875	2,628	40,734
New Mexico	132	792	170	850	709	13,471
Total cash value		$57,304		$37,320		$445,280

Source: USDA, 1959 (2).
* $6.00 per ton for corn silage.
† $5.00 per ton for sorghum silage.

TABLE 12.12. *Volume and value of harvested forages in the Great Plains, 1968*

State	Corn silage*		Sorghum silage†		Hay (all)	
	(000 tons)	*(000 $)*	*(000 tons)*	*(000 $)*	*(000 tons)*	*(000 $)*
North Dakota	2,091	$ 15,675	14	$ 88	3,779	$ 56,685
South Dakota	4,935	37,012	585	3,685	5,127	99,976
Nebraska	4,120	30,900	976	6,148	6,067	133,474
Kansas	3,388	25,410	3,717	23,417	4,741	99,561
Oklahoma	285	2,137	238	1,499	3,001	64,521
Texas	1,155	8,622	1,356	8,542	4,587	107,794
Montana	580	4,350	. . .	. . .	3,585	80,662
Wyoming	418	3,135	. . .	. . .	1,680	33,600
Colorado	3,456	25,987	275	1,732	2,858	77,166
New Mexico	450	3,375	90	567	1,049	27,798
Total cash value		$156,603		$45,678		$781,237

Source: USDA, 1969, Tables 40, 72, and 394 (3).
* $7.50 per ton for corn silage.
† $6.30 per ton for sorghum silage.

FUTURE PRODUCTION

LOOKING at the Great Plains as an agricultural resource it is possible to envision that the area will increase its agricultural income by 25 percent by 1980. This will be achieved for the most part through increased livestock production and a decrease in livestock imports from other regions. Added acreage of irrigated lands (a potential of 1.2 million acres in Nebraska alone) will result in increased harvested forages and provide pastures of high productivity over a full growing season.

There will be increased use of nitrogen and phosphorus fertilizer on semiarid and subhumid grasslands. Crop residues will be more effectively used in livestock production systems. There is a possibility that production of wheat for human consumption will be converted to livestock enterprises with wheat production for forage and concentrate feed in much of the wheat-fallow region. When these factors are combined with grazing management and weed control, livestock production increases of 50 percent or more can be expected.

SUMMARY

THE GREAT PLAINS states are uniquely suited to livestock production enterprises. This is reflected in agricultural income statistics which reveal that in 1968 the cash sales of livestock represented 64 percent of the total agricultural income of the region. Beef cattle constituted

67 percent of the cash livestock sales after these values were adjusted by subtracting the cost of livestock imported into the Great Plains states. Forages represented 33 percent of the value of adjusted income from livestock sales in 1968. Harvested forages have increased in their contribution to total forage value, and range and pasture have decreased by 8 percent in a ten-year period (1958–68). Hay represents 79 percent of the value of all harvested forages in 1968, but the major increase in production of harvested forage was represented by the 220 percent enlargement in the harvest of corn silage.

The expected increase in Great Plains agricultural income will be achieved through increased livestock production in the Plains area and a decrease in livestock imports from other regions.

REFERENCES

1. Hodgson, H. J. 1968. Importance of forages in livestock production in the U.S. Am. Soc. Agron. Spec. Publ. 13.
2. USDA. 1959. Agricultural statistics.
3. USDA. 1969. Agricultural statistics.

13

IMPORTANCE OF RANGELANDS TO THE LIVESTOCK INDUSTRY AND THE NATIONAL ECONOMY

DONALD A. PRICE

THE RANGELANDS of the United States have played an important role in the settlement and prosperity of the West. No history of the West is complete without a review of the livestock industry. Cattle and sheep ranching provided the sole sources of revenue from these rangelands in the frontier days. Today the competing uses for rangelands cause questions to be raised about their economic importance. The economy of many local areas and regions has continued to depend on the rangeland resource. Before the settlers arrived on the scene with domestic livestock, wild animals grazed these rangelands. Now for more than a century and a half they have also supported large numbers of cattle and sheep that are an important part of the economy. These rangelands are a recognized part of the outdoor scenery through the West. They are not basically suited to crop production or other intensive agriculture but can be used by domestic livestock and wildlife that can convert the forage on this vast acreage into food products useful for human consumption. Furthermore, rangelands have great watershed, open space, recreational, and aesthetic values.

THE RANGE RESOURCE

FOR THE PURPOSE of this discussion, the range resource includes all grassland pasture and range (excluding cropland pasture) and forestland pasture and range that are used or usable for grazing.

A rangeland research committee of the U.S. Forest Service (18) defined rangeland vegetation as follows:

Ranges are uncultivated areas that support herbaceous or shrubby vegetation. The range complex (ecosystem) includes not only the vegetation and soil but also the associated atmosphere, water, and animal life. Most ranges are covered with native plants, but extensive areas have been seeded to exotics. Some areas are both range and forest; the tree overstory may be sparse, or the trees may have been harvested or burned allowing growth of herbs or shrubs.

Size and Location. It has been estimated that livestock grazed about 865 million acres of land, or 38 percent of the total land area in the United States, in 1964 (Table 13.1) (6). Of this total, 640 million acres were grassland pasture and range, and 225 million acres were forestland pasture and range (6).

A major part (89 percent) of the 640 million acres of grassland pasture and range is located in the northern Great Plains, southern Great Plains, Mountain, and Pacific regions that make up the 17 western states. The 11 western states of the Mountain and Pacific regions contain 58 percent of the national grassland pasture and range. This range consists mainly of tame and native grasses and legumes but also includes shrub and brushland not classified as forest.

TABLE 13.1. *Pasture and range by type and region, United States, 1964*

Region	Cropland pasture*	Grassland pasture and range†	Forestland pasture and range‡	Total acres§	Total land area
		(000 acres)			*(%)*
Northeast	2,559	7,110	3,573	13,242	12
Lake States	4,219	8,485	6,937	19,641	16
Corn Belt	11,228	20,335	12,179	43,742	26
Northern Plains	4,159	80,675	2,300	87,134	45
Appalachian	9,457	10,778	8,356	28,591	23
Southeast	3,593	12,564	18,773	34,930	28
Delta States	4,873	9,433	27,428	41,734	45
Southern Plains	8,937	118,378	26,381	153,696	72
Mountain	4,574	314,399	85,327	404,300	74
Pacific	3,764	54,307	32,568	90,639	44
Total, 48 states	57,363	636,464	223,822	917,649	48
Alaska	4	2,772	367	3,143	1
Hawaii	52	1,203	331	1,586	38
U.S. total	57,419	640,439	224,520	922,378	41

Source: Frey et al., 1968 (6).
  * Mainly cropland in rotation, used some years for cultivated crops and other years for pasture.
  † Excludes cropland used for pasture.
  ‡ Used or usable for grazing.
  § Excludes 58 million acres in federal grazing districts and national forest system range allotments characterized by little or no use for grazing.

Sixty-five percent of the 225 million acres of forestland pasture and range is located in the 17 western states and 53 percent of the total is found in the 11 western states. The Southeast and Delta regions account for 21 percent of this forage resource. This estimate of acreage includes open forest, cutover areas, and other intermingled areas that have forage and are grazed.

Ownership. Private interests own approximately 64 percent of the 640 million acres of grassland pasture and range and 70 percent of the 225 million acres of forestland pasture and range. The remaining grazing land is held by federal, state, and local governments and as Indian lands. Of the 765 million acres owned by the federal government in the 50 states, about 273 million acres are allocated for grazing (6). The public grazing land resources are found mainly in the Mountain, Pacific, and Great Plains states. About 96 percent of all public land grazed is found within the area of the 11 western states, of which Nevada has the largest percentage of public land (87 percent) and Washington has the least (29 percent). Much has been written on the contribution of federal rangelands to the total livestock industry. This discussion will attempt to describe the total contribution of the 865 million acres of grazing land without regard for ownership.

Contribution to Total Feed Consumed. One method of ascertaining the contribution of grassland pasture and range forage to total livestock feed consumption is to obtain livestock consumption in feed units (a feed unit equals one pound of corn) by kind of feed rather than by how many and what kind of livestock were maintained on the various land areas. A total of 922 million acres of cropland pasture, grassland pasture and range, and grazed woodland contributes about 39 percent of the total feed consumed by all livestock. This is very revealing; however, of greater interest would be determination of the contribution of the 865 million acres of grassland and forestland pasture and range exclusive of cropland pasture. In addition, we need to look at the contribution of grazing to total consumption of selected classes of livestock as listed in Table 13.2. This classification excludes poultry and other livestock, which changes the picture somewhat from that usually reported.

The information in Table 13.2 brings up to date similar information presented by Upchurch in 1959 (23). From this we may see that the total contribution of cropland pasture and other pasture

TABLE 13.2. *Feed consumed by class of livestock and by source, United States, 1967*

Livestock class	Total feed consumed in the United States	Grain and concen- trates	Har- vested forages	Crop- land pasture	Other pasture and range	Total feed consumed from other pasture and range
	(000 tons)		*(%)*			*(000 tons)*
All cattle	262,750	26	26	17	31	81,453
Dairy*	87,973	33	39	10	18	15,835
Other†	174,777	22	19	21	38	66,415
Sheep and goats	12,277	8	7	30	55	6,752
Horses and mules	7,673	20	30	17	33	2,532
Hogs	56,069	96	. . .	4	. . .	. . .
Total	338,769	36	20	18	27	90,737

Note: Feed consumed is converted to feed units. A feed unit is equivalent to one pound of corn.
 * Includes milk cows and other dairy cattle.
 † Includes cattle on feed and other beef cattle.

and range is 18 and 27 percent respectively. On a regional basis the contribution of pasture and range is much greater. It was estimated from Upchurch's information that grassland pasture and range contributed to about 40 percent of livestock feed consumption in the 11 western states.

All livestock (cattle, sheep and goats, horses and mules, and hogs) consumed a total of about 677.5 billion feed units in 1967 (Table 13.2). The percentage composition of the total intake was about 36, 20, and 44 for concentrates, hay and other harvested forage, and pasture respectively. This means that approximately 288 billion feed units consumed by all livestock in 1967 came from 922 million acres of cropland pasture, grassland pasture and range, and forestland pasture and range. From these statistics we may obtain an average of 312 feed units per acre for all types of pasture; however for this discussion we are interested in how many feed units pasture and range (excluding cropland pasture) contribute.

Table 13.3 presents a summary of the projections of yields from the various types of pasture (11). Note that cropland pasture yield was estimated at 1230 feed units per acre in 1960; therefore from this yield estimate it may be calculated that cropland pasture (57 million acres) contributed 70.6 billion feed units of the total feed consumed from all pasture, leaving a balance of 217 billion feed units, which apparently must come from other pasture and range. However, the

TABLE 13.3. *Pasture yields, 1950, 1960, and median projections for 1980, 2000*

Year	Cropland pasture	Open permanent pasture on farms	Woodland pasture on farms	Grazing not on farms
		Feed units per acre		
1950	985	195	95	56
1960	1,230	220	100	60
1980	1,620	260	112	67
2000	2,120	310	125	74

Source: Landsberg, 1964 (11).

low estimate of yield per acre from open permanent pasture on farms (220), woodland pasture on farms (100), and grazing not on farms (60) does not approach that needed to supply the 217 billion feed units. The availability of feed from grazing and other than cropland was only 129 billion feed units for 1960 (11). This information seems to indicate that either the estimate of feed units per acre is too low for cropland pasture or too low for other pasture and range, or both. Another very important factor is the difference in acres of cropland used by Landsberg in his projections (79 million acres) and the amount given by Frey (57 million acres) in Table 13.1 (6). It was projected by Landsberg that by 1980 the median demand for roughage needed from grazing would be 287 billion feed units. This of course is much closer to the estimate of 287.6 billion feed units of roughage consumed by grazing animals in 1967. The assumption in the projected calculation was a yield of 1620 feed units per acre from cropland pasture and the remainder from other pasture and range. From these data we must conclude that there is a great need for more reliable information on the yields and acreages of all categories of pasture and grazing land.

Another aspect of the productivity of grazing lands is reported by the Public Land Law Review Commission (21). This group estimated the total grazing cattle population in 1966 at 66 million head and sheep and lambs at 21.5 million head. To improve the estimate of animals actually grazing, the numbers were adjusted to exclude feedlot cattle. It was also estimated that total forage consumption by grazing cattle and sheep amounted to 796 million AUMs (animal unit months) for the United States, and 164 million of this, or 21 percent, was from the 11 western states.

This same group reported that on public lands of the West there was great variation in grazing capacity ranging from 5 acres per AUM

TABLE 13.4. *Description of rangelands in the 48 contiguous states*

Type of range	Estimated average yield	Acres
	(lb/acre)	*(mil)*
Short grass	300	198.1
Sagebrush grass	400	96.5
Semidesert grass	200	89.3
Open conifer forests	700	87.0
Pinyon-juniper	300	75.7
Oak-hickory	700	57.0
Loblolly–shortleaf pine–hardwood	500	54.5
Pacific bunchgrass	700	42.5
Salt-desert shrub	200	40.9
Woodland-chaparral	500	31.0
Southern desert shrub	200	26.9
Longleaf and slash pines	1,000	26.5
Mountain grassland	900	25.0
Oak-pine	1,000	24.7
Tall grass	1,500	18.5
Alpine	700	8.0
Open aspen forests	1,500	5.0
Mountain meadows	2,000	1.0
Total		908.1

or less to 25 acres per AUM or more. This would be expected with the great diversity of lands in the West. Meadows, open parks, and gentle slopes of the mountain regions are often very productive of livestock forage. Fewer than 5 acres and often less than 1 acre may be required to support an AUM on this type of range. The semidesert ranges consisting of sagebrush, grass, and salt-desert species are generally less productive and require 10–15 acres or more of land to support an AUM.

The estimated extent of the different types of rangeland in the 48 states is presented in Table 13.4. Grasses and shrubs are predominant on 653 million acres, and 255 million acres are forested ranges (18). The average yield of the different forage types in pounds per acre (air dry) presented in Table 13.4 are from unpublished information. The average amount of forage per acre over all range types is 486 pounds, of which half (243 pounds) is estimated to be usable forage. This gives an estimate of 210.2 billion pounds of usable forage consumed from the 865 million acres of rangeland grazed. More information concerning the nutrient value of this forage is needed so that it can be converted to feed units. Probably the substitution value per pound over all forages would range from 30 to 50 percent of a pound of corn.

RANGE LIVESTOCK INVENTORY AND INCOME

In 1968 more than half (57 percent) of all beef cattle and calves in the United States were located in 17 western states (22). Beef cattle in the northern and southern Great Plains regions made up 36 percent of this total, while those in the 11 western states constituted 21 percent. The grazing cattle population in the 11 western states (adjusted to exclude feedlot cattle) increased from 11.7 to 13.6 million between 1962 and 1966. This closely followed the national trend.

During this same period the sheep population in the nation continued to decline from 26.7 to 21.5 million. Of this total about 78 percent of the sheep and lambs were in the 17 western states. According to the Public Land Law Review Commission Report (21), the 11 western states' proportionate share of the national total number of cattle did not change from 1962 to 1966, but remained at about 20 percent. The percentage of sheep and lambs in this region increased slightly, from 43.1 to 45.6 percent.

The 17 western range states that make up the major part of the nation's grassland pastures and forestland pastures and ranges contributed to more than half the livestock production and value. In 1966 the value of production from cattle was close to $8 billion nationally. The 11 western states' share in this amounted to about $1.5 billion. Value realized from sheep nationally was about $260 million, of which $120 million came from the western region.

A closer look at the importance of income from livestock to the various western states reveals that in 1966 cattle and calves were the major source of cash receipts to farm income in 15 of the 17 western states. The two exceptions were North Dakota and Washington, where wheat was the leading commodity for cash receipts. Cash receipts from cattle and calves ranked second and third for North Dakota and Washington respectively. Sheep and lambs also were in the top five categories for cash receipts in this region.

It was estimated that 60 and 55 percent of the agricultural income in Nevada and Wyoming respectively was dependent upon range and pastureland, while California and Washington showed a lower dependence of 15 and 10 percent respectively (23). An estimated $30.1 million of annual income to Wyoming's sheep and wool industry generated approximately $71.6 million of commercial business; local, state, and federal government revenue; and personal income for the state (17).

Nevada researchers found that in 1967 the meat animal industry in Nevada generated $77.4 million for the state economy, range forage contributed approximately 52 percent of the feed consumed by the livestock, and the remainder was supplied by irrigated lands (12).

What will be the effects on the rural economy of reducing rangeland grazing? It has been estimated that a 20 percent reduction in federal grazing would cause an 11 percent decrease in gross ranch income (3). To simulate the effect of this reduction in Oregon, all ranches dependent on federal range were studied in one Oregon county (2). The dependent ranches generated over $3 million worth of new money in the county from export sales and spent almost $1.8 million for goods and services, which included such items as agricultural, automotive, construction, and transportation services. As a result of economic interdependence in this rural county a reduction in rangelands caused an 11 percent reduction in ranch income and $623,342 total reduction in county business receipts.

Since production of beef cattle and sheep in the range states involves approximately the same combination of inputs for services required from the business community, these studies give some insight into the importance of federal grazing lands. The importance of all other rangelands would be even more significant. More of this kind of information on interindustry dependence would be useful for evaluating the true importance of rangeland resources to the livestock industry and the economy of the nation.

RANGELANDS AND MULTIPLE USE

The federal government requires that land administered by federal agencies be managed for multiple use such as timber, water, recreation, wildlife, and forage. This requires that the livestock industry review its grazing management system to adapt to these other uses.

Until recently many people believed that the range was the exclusive property of the domestic livestock industry and that forage could be converted into economic uses only by grazing animals. However, today in many areas ranchers are being subjected to ever-increasing competition from recreation, wildlife, and other uses that must be accommodated, especially on public rangelands.

A study in the Paunsaugunt area of Utah (15) estimated the dollar value of five land resources taken at the point of harvest. Recreation, wildlife, water, livestock forage, and timber ranked in importance in the order listed. As the resources moved into the economy the rank was recreation, water, wildlife, timber, and livestock forage. Recreation had an estimated dollar value expanded into the economy of $2.8 compared to $0.4 million for livestock forage. Range forage, the value of which is measured through the livestock consuming it, turned out to have the lowest dollar value of the five resources in this multiple-use area. However, 75 percent of the estimated $0.4 million value for range forage remained inside the local rural economy.

From the standpoint of dollar value only, the range livestock industry was in a relatively weak position in the Paunsaugunt area. However, when the competitive situation between cattle and deer was examined it was found that even though deer generated more wealth through sale of hunting privileges, they were held to be of little value to the local community, whereas livestock values had great local significance. Somewhat different findings were reported from Nevada. In 1967 the meat animal industry in Nevada generated $77.4 million in the state economy; big-game hunting contributed a much lesser amount ($4.6 million) in the same year.

There are complementary effects from multiple use by wildlife, timber, recreation, and other uses with range livestock. In Nevada competitive uses of the range forage by livestock and big game was studied (12). After evaluating past studies on the grazing habits of cattle and big game they concluded that on brush-grass ranges, the big game strongly preferred browse, while cattle consumed mostly grass. It appeared that deer and cattle do not consume exactly the same browse species; 50 percent of the browse species consumed by deer have yet to be found in the diets of grazing cattle, and 25 percent of the deer diet consists of browse species only occasionally taken by cattle (12). The main point, then, is that livestock grazing reduces grass and allows an increase in shrubs for big game. Finally, it was concluded that Nevada's rangelands should be utilized by both the grass- and browse-consuming species in order to maintain a forage habitat that would be beneficial for both deer and cattle. Removal of livestock from Nevada ranges would eventually lead to a reduction in the deer population through excessive use of browse species.

The timber industry also can benefit from grazing by range livestock. Returns from grazing may under certain conditions equalize or even exceed any damage that may have resulted to young trees by aiding reproduction and reducing fire hazards. In the western states, where most coniferous forests are found, such damage to reproduction of the trees by livestock grazing is seldom serious where proper range management is practiced (19). Grazing livestock increased reproduction in the northwestern Douglas fir forests by dropping seed and by decreasing competition from herbaceous plants (10). Possibly the greatest benefit to the timber industry derived from livestock grazing comes from the reduction of the fire hazard by removing inflammable vegetation (20). In addition, livestock trails and driveways have served effectively as fire lines.

The livestock industry in turn can benefit from the activities of the timber industry. Forest Service researchers in Montana (1) found that clear-cutting of lodgepole pine stimulates production of understory vegetation that may provide a grazing resource for livestock and big game for an estimated 20 years or more. Peak production of 800–1000 pounds per acre occurred about 11 years after clear-cutting.

Recreation and the livestock industry have the potential of being the most complementary of the various range resource uses. Dude ranches are becoming more popular every year, and owners of western rangelands who manage properly can support both livestock and big game, giving the dude ranch enterprise an advantage in the range states over other areas. As our cities become more crowded and the pressures of day-to-day business life become more intolerable, the open landscapes with grazing cattle, sheep, and big game will be subjected to more recreational use and be valued for their aesthetic qualities. The livestock producers, the federal land managers, and the recreationists are becoming more aware of the importance of rangelands.

Water production is complementary to grazing livestock unless grazing is improperly managed. If grazing contributes to watershed deterioration, it is competitive with water for irrigation and urban uses. Water production for direct use by livestock and irrigation of livestock forage crops is also of complementary value to the livestock industry.

IMPORTANCE OF RANGELANDS
TO THE FOOD SUPPLY

THE DEMAND for rangeland forage will increase if the predictions of future human demands for grains are at all accurate. In this country up to 1972 with the great grain surplus of that period there was an economic advantage for feeding grain. To meet the requirements of a rapidly expanding population, our long-range thinking must be concerned not only with the relative advantages of grain feeding compared to forage feeding but especially with producing red meat from forage with a minimum of grain.

In many instances livestock grazing cannot compete with other higher use requirements such as military reservations, atomic energy projects, expanding cities, highways, and other demands. This situation will continue as long as this nation has an abundant food supply. However, when meat supplies become short, the demand for meat products will increase and grazing may then be competitive with other uses.

Sparsely populated areas such as the western range states and other minor grassland areas in the United States will continue to be major sources of livestock products, and increased productivity will result through improved range and livestock management. We must expect, however, that any rangelands that have potential for farm crop production will be cultivated and eventually will not be available for grazing. Other uses of rangelands will undoubtedly increase too. The extent to which these lands will be plowed depends largely on the rate of population growth.

In 1959 the United States had about 2.53 acres of arable land and 3.49 acres of grassland per person. When our population reaches 400 million, the amount of arable land per person will be about 1.14 acres and the amount of grassland about 1.58 acres, both below the 1959 world averages of 1.18 and 2.16 acres. The total agricultural land area per person in 1959 was about 6.0 acres. This decreased to about 5.5 acres in 1967, or an average of 1 percent per year (14).

The estimated population in the United States will be 330 million in the year 2000. The per capita consumption of beef, veal, lamb, and mutton was 113.1 pounds in the United States in 1967 or about 22.5 billion pounds. If we assumed that per capita consumption would remain the same, the demand for beef, veal, lamb, and mutton would be 37.3 billion pounds in 2000. The per capita consumption of beef has almost doubled since 1940. Based on this trend, the livestock industry by the year 2000 may have to produce at least 66 percent more beef, veal, lamb, and mutton than it did in 1967 or this country will have to import these meat products or consume relatively more pork and poultry products if they are available. This is indeed a challenge. However, further increases in efficiency of production of livestock, grains, forage crops, and better range and pasture management should more than compensate for increased population and loss of land for other uses. It should also be mentioned that, because of food surpluses prior to 1973, this country had idle reserves of agricultural land which could have been used for increased production.

BENEFITS FROM ANIMAL AND
RANGE SCIENCE RESEARCH

HIGHER PRICES and more efficient production of livestock products are essential to motivate the livestock industry. Therefore animal and science research programs must be given the necessary support to provide the improved technology for increasing animal and range forage productivity.

There are many ways rangelands can be managed and improved to increase productivity. Some of the most successful methods are improved management, artificial reseeding, fertilization, and eradication of noxious vegetation. Much more research needs to be conducted and applied to management of livestock on the range, to measuring range condition and trends to vegetation and soil surveys, and to management planning.

Introduced grasses such as crested and intermediate wheatgrasses are well adapted to depleted ranges and can be grazed earlier in the spring than most native grass species. On foothill ranges in Utah 12.4 acres were required per AUM before seeding but only 3.1 acres after seeding. Steers and calves gained almost a pound per day more on seeded ranges than on adjacent native ranges. Over a ten-year period on the same seeded foothill ranges lambs gained 0.06 pound per day more (5).

There are an estimated 330 million acres of brush-infested rangelands in the United States and approximately 96 million acres of western range in sagebrush, which in tall dense stands is definitely undesirable. It is relatively unpalatable to sheep and cattle and uses moisture and nutrients that should be producing better forage. Numerous instances have been reported where sagebrush eradication more than repaid its cost on many western ranges; from double to 25 times greater capacities have been obtained. Sagebrush burning experiments in Idaho increased grazing capacity an average of 69 percent. Perennial grasses and forbs increased 60 percent, partially replacing the sagebrush (13). In Colorado, Hanson (7) obtained a 300 percent increase in grazing capacity by burning sagebrush range. Similar increases have resulted from spraying sagebrush with herbicides. This has improved grass yields as much as 1500 pounds per acre (5). Increased livestock production from brush-burning is also realized by easier handling of sheep and cattle on sagebrush burns where they graze with less difficulty. Lamb losses from straying and from predators are greatly decreased.

Fertilization of rangelands has had limited application because research has shown conflicting results. One study showed that the addition of 30 pounds of nitrogen per acre on mountain ranges increased forage from 400 to 2000 pounds per acre (5). Another study (8) indicated no increase from application of nitrogen. On seeded foothill ranges in Utah, increased forage yields averaging 1300 pounds per acre have resulted from the addition of 30 pounds of nitrogen per acre (5). Fertilization not only increased the quantity of the forage but the nutritive value and vigor of the range plants. On California annual ranges, addition of sulfur (either alone or in combination with nitrogen) usually stimulated growth (4). However, in many studies on rangeland, increase in production is not sufficient to justify the cost of fertilizer. Much more research is needed to develop the best times and rates of application.

Fencing is an aid in the management and development of rangelands. Where it is practical, fencing is gaining wide acceptance on lands of lower productivity. Better market prices will make the economics of fencing more favorable. Fencing has many advantages; it eliminates need for herders, and research has shown as much as 25 percent increase in grazing capacity. Fencing permits common use by cattle and sheep, offers an opportunity to practice deferred and rotational grazing to rest pastures, and results in heavier calves and lambs at market time (16). Fencing permits loose lambing and calving, which disturbs the ewe or cow less so that offspring obtain a more even supply of milk and, in the case of sheep, the mother will be less likely to leave a twin behind. Clean wool yield has increased as much as 6–8 percent with sheep under fencing. Obviously, the main disadvantage of fencing is cost. Losses from predators may also increase under fencing, especially with sheep; and new water developments are necessary.

Water at frequent intervals is one of the most important needs for increasing productivity of livestock on western ranges. When water is provided to all parts of the range, the entire area can be uniformly and properly grazed. When permanent watering places are used, the range is overused along the trails and loss of sheep or cattle from grazing poisonous plants is increased. It was reported on one range operation that prior to hauling water the lamb crops averaged 80 percent and weighed 60 pounds at market time in the fall. After including water hauling in the management scheme, lamb crops averaged 110 percent and market weights, 72 pounds. In another trial, sheep supplied with water every day on desert winter range gained

3.4 pounds per head in 40 days; those receiving water every second day gained 0.8 pounds; and those watered only every third day lost 6.0 pounds per head (9).

One of the major obstacles to efficient utilization of rangelands is lack of adequate knowledge of their present condition and potential productivity. The agencies charged with consultation or with management of public and private rangelands are working with methods to ascertain utilization and condition and methods of inventory that are not adequately based on research results. Additional funding should be provided for research to allow more precise decisions in future range management.

The application of research has advanced production in practically all aspects of animal industry. There are still other potential means of increasing production without greatly increasing animal numbers. Domestic livestock can be managed and improved to produce more efficiently in many ways. Research information in animal breeding, reproductive physiology, wool technology, nutrition, and management is available. Historically, there has been a considerable lag from the time research information has become suitable for practical application by the livestock industry and the time it is actually put to use. Some of the better known production methods now available are crossbreeding, performance testing, estrus synchronization, artificial insemination, and pregnancy diagnosis. New methods of increasing the number of births per cow and sheep are fast reaching the realm of practical application.

Increased research emphasis in range nutrition and management will pay greater dividends in the future. For example, the use of nonprotein nitrogen compounds as a replacement for supplemental protein can provide for more economical livestock production.

Control of diseases and parasites has improved markedly through research and application of antibiotics, sulfas, and anthelmintics. The importance of livestock sanitation to disease prevention is beginning to be appreciated by livestock producers.

Increased efforts in promoting greater product use and expanded markets, especially in the sheep industry, must receive more attention and support to make lamb and wool more competitive with other livestock products and synthetic fibers and cotton.

SUMMARY

EVEN THOUGH rangeland grazing will continue to be an important part of the economy of the livestock industry, the animal scientist as well as the range scientist must attack research problem areas from the range ecosystem concept. The value of rangeland is not only the forage but also the water, timber, minerals, recreation, and wildlife. These resources must be maintained along with the quality of the environment of our rangelands. More research needs to be done to determine the impact of all uses of the range on the soil, water, air, fish, and other wildlife.

Competition between animal species will continue and the more efficient users of concentrated feeds will cause shifts in relative amounts of animal products produced from the various species. However, the range livestock industry has a bright future in that only ruminants can convert the vast resources of range forage into animal protein for human consumption.

REFERENCES

1. Basile, J. V., and C. E. Jensen. 1971. Grazing potential on lodgepole pine clearcuts in Montana. USDA Forest Serv. Res. Paper INT-98.

2. Bromley, D. W. 1968. Economic importance of federal grazing on interindustry analysis. Oreg. Agr. Exp. Sta. Tech. Paper 2440.

3. Caton, D. D. 1965. Western livestock ranching and federal rangelands. ERS, USDA and USDI.

4. Conrad, C. E., E. J. Woolfolk, and D. A. Duncan. 1966. Fertilization and management implications on California's annual-plant range. *J. Range Manage.* 19:20–26.

5. Cook, C. W., and L. A. Stoddart. 1964. Range resources—Intensive management can meet future demand. *Utah Farm and Home Sci.* 25 (4): 100.

6. Frey, H. Thomas, Orville E. Kraus, and Clifford Dickason. 1968. Major uses of land and water in the United States, with special reference to agriculture—Summary for 1964. USDA Agr. Econ. Rept. 149.

7. Hanson, H. C. 1929. Improvement of sagebrush range in Colorado. Colo. Agr. Exp. Sta. Bull. 365.

8. Hull, A. C., Jr. 1963. Fertilization of seeded grasses on mountainous rangelands in northeastern Utah and southeastern Idaho. *J. Range Manage.* 16:306–10.

9. Hutchings, S. S. 1958. Hauling water to sheep on western ranges. USDA. Leafl. 423.

10. Ingram, D. C. 1931. Vegetative changes and grazing use of Douglas fir cutover land. *J. Agr. Res.* 43 (5): 387–417.

11. Landsberg, H. H. 1964. Natural Resources for U.S. Growth. Johns Hopkins Press, Baltimore. (Publication for Resources for the Future, Inc.)

12. Lesperance, A. L., and P. T. Tueller. 1969. Competitive uses of Nevada's range forage by livestock and big game. RNR Rept., Renewable Resource Center, Univ. Nevada.

13. Pechanec, J. F., A. P. Plummer, J. H. Robertson, and A. C. Hull, Jr. 1965. Sagebrush control on rangelands. USDA Agr. Handbook 277.

14. Phillips, R. W. 1963. Animal products in the diets of present and future world populations. *J. Anim. Sci.* 22:251–62.

15. Ridd, M. K. 1964. An area-oriented approach to multiple land use research. Economic research in the use and development of range resources. USDA Rept. 6.

16. Roberts, W. P., Jr. 1961. Fencing versus herding of range sheep. Univ. Wyo. Agr. Exp. Sta. Circ. 156 (mimeo).

17. Roehrkasse, G. P. 1962. The economic value of the Wyoming sheep and wool industry to Wyoming's economy. Univ. Wyo. Agr. Exp. Sta. Circ. 168 (mimeo).

18. Rummell, Robert S. 1970. Range ecosystem research: The challenge of change. USDA Agr. Inf. Bull. 346.

19. Sampson, A. W. 1962. Range Management Principles and Practices, 3rd ed. Wiley, New York.

20. Shepherd, W. O., C. M. Kaufman, and H. H. Biswell. 1946. Forest grazing in North Carolina. *Southern Lumberman* 173:228–38.

21. University of Idaho, with Pacific Consultants, Inc. 1970. The forage resource. Study for the Public Land Law Review Commission.

22. USDA. 1968. Livestock and poultry inventory. Stat. Rept. Serv. annual summaries.

23. Upchurch, M. L. 1963. Public grazing lands in the economy of the West. *In* Land and Water Use, pp. 83–98. Am. Assoc. Adv. Sci. Publ. 73.

14

I.   THE RANCHER AND OUR NATIONAL FOREST AND GRASSLANDS

ROBERT S. RUMMELL

"CARL, time you get the horses, breakfast will be ready." That command from his father was the start of a day momentous to 15-year-old Carl Franklin. For this was June 1, the day the Franklins' cattle were to be put on the national forest for a summer of grazing.

Carl and his father had started their herd up the trail from the home ranch three days before, and today would complete the trail drive. The district forest ranger was due to be at the Shake Creek Gap at eight o'clock to count the cattle in. The Franklins would push their herd up Meadow Creek and out onto the broad bunchgrass slopes of the Flat Ridge cattle allotment where the cows and their young calves would rest up and then begin their summer foraging.

Carl remembers well the brown saddle blanket that almost kept off the chill as he stared up from his bough bed at the dawning east. "Well, let's go. Those old cows sure know where we're a-headin'. They'll be at that gate all by themselves afore long, I reckon."

As Carl and his father rode behind the red and black and brown cattle, now bawling as they sensed the closeness of their summer home, Carl thought with pleasure of what lay ahead. This was to be his first season to stay alone with the herd. He would put up at the cow camp.

There would be wood to cut, his own meals to cook, and horses to wrangle. He would pack blocks of salt where the cows would find and lick them into rounded mountains and valleys, and then into blobs of nothingness. He would ride fence and ax-free the winter-fallen lodgepoles, and splice and stretch and staple the sharply barbed wire.

Reprinted with permission from 1967 USDA Yearbook Agr. (1).

There would be other tasks. So many he would have but little time to search out the den of coyote pups yammering away at the moon. Or see if Old Crip, the three-legged bear, was still around.

But it takes a man to punch cows. Carl was 15. And a man. So there was a job to do. That's the way it is.

YEARS PASSED and now the herd was Carl's. It was another summer, and Carl took the cattle to the forest for grazing.

"Well, Mr. Franklin, all the 115 head counted in. Your permitted number right down to the last set of horns." District Ranger John Simmons reminded Carl of that other ranger who had met him and his father at the Shake Creek Gap that other June 1. The summer Carl first spent alone with the Franklin herd. "Your cows wintered real good! They'll stay sleek-coated this summer. Your range sure has improved under that new management plan we worked out."

John Simmons had come to the Flat Ridge District two summers ago. In this short time, he had become acquainted with the 200,000 acres of forest and rangeland in his district and its problems. He had traveled from the dry foothills, up through the easy-laying pine forest, on through the higher timbered slopes, and up into the subalpine grassland. He knew all the permittees and their home ranch locations and grazing practices. He had discussed range management with Carl, and he had ridden the range with him.

"Putting that fence across Battle Flat has turned the trick. That and the new watering pond have stopped overuse of the grass. Your cows don't concentrate on the meadows anymore."

That fence and the watering pond were a part of the plan for managing livestock on the Flat Ridge allotment under multiple use, for Carl Franklin and District Ranger Simmons had spent many hours discussing how to develop and manage the range resource on Flat Ridge. They wanted enough forage for Carl's cattle. At the same time, they tried to consider all the needs and requirements of the land and its users. They wanted to maintain a good grass cover for watershed protection. Carl and John knew that the elk which foraged on the higher reaches of the allotment would require grass and sedges, so they reserved some of the allotment's grazing capacity for big game.

Their plan for restoring the grass cover on Battle Flat, where the good forage plants had been replaced by sagebrush, was to spray-kill only part of the sage; and the rest was to be left for sage-grouse cover and food. Because increasing numbers of picnickers used the aspens at Shady Grove, the plan was adjusted so the cows grazed there only after Labor Day. This left the grove and its picnic area free of cattle during most of the picnicking season.

"I believe we've got her coming our way, John. The grass is strong and getting better. I'll be back to push 'em through the gate again next year."

So it had gone each opening day of the grazing season for the Franklins and for the thousands of other cattle ranchers over the years. And for the sheepmen, too, with their bands of ewes and lambs moving into national forest system lands.

Since 1905, livestock have grazed the national forests under permit. Ranchers owning livestock and land apply for the privilege of grazing their livestock. If they qualify and the grazing is available, they are then issued a permit. And there is a charge made for this grazing.

Although the first permits were issued in the western states, grazing now is permitted on the national forest system lands in 39 states. Only in nine eastern seaboard states and Kentucky and Hawaii is there no livestock grazing under permit.

It is a big spread, this national forest system, with over 100 million acres divided into 11,600 grazing allotments and grazed by 7 million cattle, sheep, and horses. (See Fig. 14.1.) It has more than 50,000 miles of fence and 38,000 livestock watering developments. Included are 4 million acres of the national grasslands, largely in the Great Plains states, and another 4 million acres of the intermingled or adjacent private land for which administration of grazing is waived to the Forest Service.

To the Carl Franklins and to the 20,000 other Franklins and Joneses and Rodriguezes who graze their cattle and sheep on the national forests, the forage from these public lands is an important part of their total ranching operation.

Over the years Carl Franklin has built up the productivity of his home ranch by seeding grasses on the native range, by using his range properly, and by practicing good irrigation on his meadowland. But even with this effort, Carl still finds that summer is a critical forage supply period, and this is where his national forest grazing permit helps. The grass in the mountains is green and nourishing in the summer, and his cows give rich milk for their sucking calves.

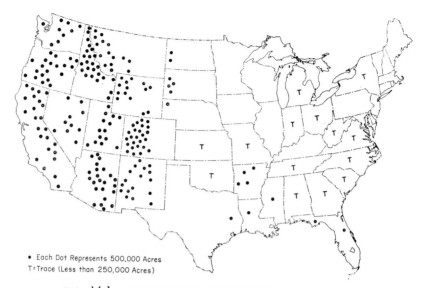

FIG. 14.1. DISTRIBUTION OF NATIONAL FOREST SYSTEM LAND
GRAZED BY LIVESTOCK, 1969.

While his brood cows summer on the forest, he grows hay on his
home ranch meadows for winter feeding. His operation is in balance.

Carl's grazing fees are set to return fair market value to the federal government for his use of the public lands. A portion of these
fees is returned to his county for aid in financing schools and roads.
In enabling Franklin and other ranchers to realize the full production value of their privately owned lands, the national forest system
promotes community stability. These ranches are the taxpaying, income-producing businesses that are the foundation of the community.

Two contrasting localities show how the national forests fit into
local livestock ranching economies. In northeastern California, 147
ranchers graze 19,000 head of cattle on the Modoc National Forest.
This is around 25 percent of the cattle in Modoc County. The ranchers want to graze more cattle, but the ranges presently are fully
stocked. As additional grazing becomes available through range improvements and development, the ranchers will have a chance to
use it. Average permit in this area is for 130 cattle with the average
operator grazing about 30 percent of his livestock on the forest range.

The second locality is in the mountains of northern New Mexico.
Here live many small stockmen, largely of Spanish-American origin,

whose families have grazed livestock on the national forest for many years. In this dry but beautiful country, 425 permittees graze 10,000 cattle on the Santa Fe National Forest. Ownership of cattle averages 35 head per rancher with the average cattle permit 24 head.

As on the Modoc National Forest, demand is high for the grazing from the Santa Fe National Forest. The stockmen depend very greatly upon this grazing, too. Of the 425 Santa Fe permittees, 198 own only an acre of land each; another 172 own from 2 to 40 acres; only 38 own more than 40 acres with 17 owning no land. And instead of a 3- to 4-month grazing season, these ranchers graze their livestock on the forest an average of 6 months a year.

In these two widely contrasting ranching localities, national forest system lands play important roles. In both, ranchers need and depend upon the national forest for livestock grazing. This grazing helps sustain the agricultural way of life and its economy for the communities where these ranchers live.

Yes, Carl Franklin knew that he would be back next year to graze his cattle on the national forest. Just as he had come back each June for 35 years. Even though productivity of his home ranch has increased through good range management and he has a much more efficient operation, Carl still needs the summer grazing from the Flat Ridge allotment. Carl's ranch today supports more cattle than it did 35 years ago, and they are better and bigger cattle at that.

On his mountain grazing allotment, the productivity is also greater. Instead of the one hewn-log trough at Punch Spring that watered cows from a square mile of range, Carl now has a pipeline that feeds three different troughs. Where the Flat Ridge once was one big pasture, it now is fenced into four units. Rotation of the cattle between pasture units on the Flat Ridge takes more time than just turning them in at Shake Creek and picking them up as they drift back there by themselves in the fall, but this lets Carl keep an eye on his quality cattle. They do very well under rotation, and the Flat Ridge range is in good condition.

And all this is the measure of District Ranger John Simmons's and Rancher Carl Franklin's success as a cooperating team of public land administrator and private user. Cattle will keep on grazing the Flat Ridge allotment as an important use of these national forest lands. This piece of the national forest is contributing its bit to the public good.

CHAPTER 14

II.   RANGE ECOSYSTEM RESEARCH:
THE CHALLENGE OF CHANGE

A CHARACTERISTIC of our times is change, and change brings
with it challenges, opportunities, and problems. Half the United
States is rangeland, and as population, industrialization, and urbaniza-
tion increase, rangeland's role expands. The economy of many localities and regions depends on range resources. In meeting the challenge
of the future and adapting to changes in society's needs, what is the
best way to proceed with research on the problems of managing, conserving, and using rangelands?

We will outline the direction the Forest Service has chosen to
take with rangeland research and explain the underlying philosophy.
Our decisions are based on the findings of a committee appointed in
1968 by the chief of the Forest Service: first, to make a critical review
of range research, second, to consider society's changing needs for
range resources, and third, to develop a comprehensive basis for research programs to meet those needs.

Society's needs and priorities are changing, and we need to take
a new look at our concepts of rangelands. To replace the traditional
concept based on land use, the following ecological concepts are proposed:

1. Ranges are uncultivated areas that support herbaceous or shrubby
    vegetation.
2. The range complex (ecosystem) includes not only the vegetation
    and soil, but also the associated atmosphere, water, and animal
    life.
3. Most ranges are covered with native plants, but extensive areas
    have been seeded to exotics.
4. Some areas are both range and forest. The tree overstory may be
    sparse, or the trees may have been harvested or burned, allowing
    growth of herbs or shrubs.
5. Rangelands have many uses and yield a variety of products. Live-
    stock and game have been the most prominent, but other outputs
    are increasing in importance and are produced even by rangelands
    that are not grazed.

Rangelands make up half of our nation's land area and are an
important part of our total environment. The widespread, diversified

Reprinted with permission from the USDA, Forest Serv. (2).

range resources include grassland prairies, brushlands, understory beneath open forests, desert areas, and other types.

Rangelands played a vital role in the economic development of the West; early use of western ranges was unregulated, and pioneer stockmen generally knew little about the ecosystems. Overuse of the ranges damaged vegetation, soil, streams and lakes, and wildlife habitats. Range research and management have helped to prevent new abuses and to repair damaged resources. The livelihood of many communities throughout the West, in the Ozarks, and in the South depends on rangeland.

Rangelands will become increasingly more important to society for various products and uses in addition to livestock. From a nationwide point of view, range livestock production is a modest part of total agriculture. Agriculture has become more specialized and more efficient, and crop and livestock production has shifted among the regions of the nation. Western ranges have lost their comparative advantages for livestock production, while advantages have increased in the Ozarks and the South. America has millions of acres of surplus cropland on which livestock may be produced. Farm pastures and croplands will meet most of the future needs for additional livestock, although some of the future supplies may come from ranges, especially in the South. Rangeland grazing, however, will continue to be an important part of the economic fabric of many localities and regions. Also, as sources of open space, natural beauty, recreation, water, clean air, and fish and wildlife, range values will increase.

Range ecosystems are composed of communities of plants and animals in their natural environment. Ecosystem components (vegetation, soil, water, air, fire, animals, topography, temperature, solar energy, and man) are closely and completely interrelated, and any influence on one affects others. Understanding the ecosystem is basic to all man's goals for use of the natural resources.

Range ecosystems are important sources of a variety of outputs, including water, air, fish and wildlife, livestock, recreation, landscape and open space, miscellaneous other products, and a resource reserve. They can contribute to present and future generations through the following societal goals: maintaining environmental quality, strengthening rural communities, and providing recreational opportunities. We have a continuing obligation and privilege to efficiently manage and conserve the rangelands.

Livestock grazing will continue as one use of the land. Range livestock production—as the major economic base of many families, communities, and regions—can contribute to the economic stability and growth of rural areas.

Maintaining environmental quality in the rangeland includes conserving soil and water, maintaining range vegetation and open spaces as sources of oxygen and clean air, and preserving the quality of landscapes and scenery.

Rangeland areas, as sources of fish and wildlife, can produce more with proper management and investment. These lands, however, provide many recreational opportunities in addition to those based on wildlife and fish.

A comprehensive program is needed to develop information and data necessary for man to properly care for and use rangelands. Major research needs include: analysis of ecosystems, inventories of range resources, coordination of management and use, improvement of resources, maintenance and improvement of environmental quality, and analysis of social and economic aspects of resource use.

REFERENCES

1. Rummell, Robert S. 1967. The rancher and our national forest and grasslands. *In* Outdoors, U.S.A. USDA Yearbook Agr.

2. ———. 1970. Range ecosystem research: The challenge of change. USDA Agr. Inf. Bull. 346.

15

IMPORTANCE OF
PUBLIC DOMAIN GRAZING LAND
ADMINISTERED BY THE
BUREAU OF LAND MANAGEMENT,
U.S. DEPARTMENT OF THE INTERIOR

GEORGE D. LEA

THE TAYLOR GRAZING ACT, which regulates grazing on U.S. federal lands, was passed in 1934 to stop injury to forage resources by preventing overgrazing and soil deterioration and to provide for the orderly use, improvement, and development of public grazing lands as well as to stabilize the livestock industry dependent upon the public range. Section 3 of the act provides for the issuance of permits to bona fide residents and stockowners to graze livestock within grazing districts. Preferences are established and permits are issued for a period of not more than 10 years. Section 15 of the act provides for the issuance of grazing leases to qualified applicants for public lands outside authorized grazing districts.

GRAZING LAND UNDER THE
BUREAU OF LAND MANAGEMENT

ABOUT 150 million acres (of a total of 174 million acres of public lands) are administered by the BLM and lie in the 11 contiguous western states where livestock grazing is important. Approximately 4.9 million animal units of all classes of livestock were grazed in 1967 by 28,800 permittees (3.7 million cattle and 6.1 million sheep). Grazing permits on federal lands attached to qualified private land or to water-base properties have been recognized for more than 30 years. Private ranch operations vary considerably in size and in terms of their dependency upon public land forage. In many areas of the Southwest, ranchers utilize the public land year-round, while in parts

185

TABLE 15.1. *Condition and trend on BLM-administered land*

Condition	Thousand acres	Percent of total	Trend	Thousand acres	Percent of total
Excellent	1,600	1	improving	24,000	15
Good	24,000	15	static/indefinite	113,000	71
Fair	84,800	53	declining	22,400	14
Poor	41,600	26			
Bad	8,000	5			

of the West, as little as one month's use is made. In many cases, public land is a vital part of the total livestock operation.

LAND CONDITION

In 1963, range conditions of BLM-administered lands were estimated as shown in Table 15.1. These are primarily arid or semiarid grasslands, and they are contributing 554 million tons of silt to the pollution of western streams; and rapid runoff rates cause over $14 million flood damage each year. See Table 15.2.

IMPORTANCE OF PUBLIC LAND FORAGE

The following discussion is adapted from Upchurch (1). Some of the data have been updated and his "index of dependence" criterion has been extended to BLM lands. The continued availability of forage from public lands is necessary to sustain the yearly economic operation of dependent individual operators. From a national dependency standpoint, the federal range produces less than 2 percent of the livestock feed consumed annually in the United States. Although little of our national supply of livestock feed comes from federal grazing lands, this does not provide the full story of the con-

TABLE 15.2. *Summary erosion profile, public domain watersheds*

Item	Units	1969 situation
	(mil)	
Area administered	acres	175.2
Area deteriorated	acres	123.0
Critically eroded	acres	45.0
Moderately eroded	acres	78.0
Annual sediment yield	tons	553.9
Flood damage	dollars	14.1

TABLE 15.3. *Feed consumed by class of livestock and by source, 11 western states,* 1959

Livestock class	Total feed units consumed in the U.S.*	Grain and concen- trates	Harvested forages	Cropland pasture	Other pasture and range
	(000 tons)		*(%)*		
All cattle (total)	32,300	18	30	17	35
Dairy	7,800	17	48	32	3
Other	24,500	18	24	12	46
Sheep and goats	5,360	3	14	16	67
Horses	1,260	11	30	7	52
Hogs	885	96	...	4	...

Livestock class	Total feed units consumed from other pasture and range*	Private land	National forests†	Grazing districts‡	Indian land	State land
	(000 tons)			*(%)*		
All cattle (total)	11,439	56	11	23	6	5
Dairy	219	100	...	...	...	...
Other	11,220	55	11	23	6	5
Sheep and goats	3,591	64	7	25	2	2
Horses	656	93	1	4	1	1
Hogs	...	...	...	...	...	...

Source: Estimates from unpublished data by Ross Baumann and Earl Hodges, Farm Econ. Div., ERS, USDA.

* Quantities in terms of feed units. A feed unit is equivalent to 1 pound of corn.

† Indicates national forests, national grasslands, land-utilization projects.

‡ Includes Taylor grazing districts and section 15 leases.

tribution of public-land forage to the range livestock industry in the West.

A regional view gives a different picture. Most of the federal rangelands are in the 11 western states. In the West, about two-fifths of all livestock feed comes from pasture and range other than cropland, and of this, about one-third comes from federal range (see Table 15.3). On the basis of volume of feed alone, the federal range hardly looms large even in the West, since it produces about one-eighth of the livestock feed consumed in the region. But again, this does not tell the full story.

The acreage of public grazing lands is unevenly distributed over the West. In four western states (Arizona, Nevada, Idaho, and Utah) more than half of the land grazed is in public ownership. In three

TABLE 15.4. *Livestock industry dependency*

State	Industry dependency by community				Community dependence			
	0–10%	10–40%	40–70%	70–100%	0–3%	3–8%	8–15%	15–20%
Arizona	5	3	...	1	6	2	1	...
California	1	1	...	1	1	2	...	...
Colorado	10	4	...	...	13	1	...	...
Idaho	7	6	...	...	11	1	1	...
Montana	4	6	...	...	1	5	3	1
Nevada	2	4	1	3	2	2	5	1
New Mexico	1	10	6	...	8	8	2	...
Oregon	3	4	...	...	4	2	1	...
Utah	2	5	3	2	4	5	1	2
Wyoming	4	7	...	1	4	6	2	...
Total	39	50	10	8	54	33	16	4

Assumptions of Public Forage Dependency

Livestock industry dependence	Community dependence	Dependency level
(%)	(%)	
0–10	0–3	insignificant
10–40	3–8	moderately significant
40–70	8–15	highly significant
70–100	15–20	critically significant

Note: Total of 107 communities studied.

others (New Mexico, Wyoming, and Oregon), the proportion of public to private grazing land is nearly 50/50. In the other four (California, Washington, Montana, and Colorado), the acreage of public grazing land ranges from less than one-fifth to about two-fifths of all land grazed.

The data presented in Table 15.4 from studies of 107 communities provide indications of the importance of public land forage to the livestock industry and to the communities themselves.

IMPORTANCE OF LIVESTOCK USE TO THE ECOSYSTEM

THE RANGE-WATERSHED complex is made up of parts so closely interrelated and normally so well integrated or adjusted to one another that what affects one affects all. The ultimate parts of the range-watershed complex may be grouped generally into five classes—animals, vegetation, soil, climate, and topography. A basic knowledge of each of these components is necessary to understand the ecology of the whole. Retainment of the soil base is prerequisite to maintenance of all the surface resources. The one factor in this ecosystem over which man has the most control is the animal.

Controlled livestock grazing is in some instances a prerequisite to maintaining game habitats in good condition. Grazing of the grass cover by livestock is necessary to maintain the ecological balance. For example, many highly desirable browse ranges would be crowded out by other vegetation without use by domestic animals. Livestock, with a preference for grass in the spring, can effectively utilize such areas early to reduce the competition from grass and provide better growing conditions for browse. By removing livestock soon enough, full vigorous shrub growth is available for wildlife in the fall and winter. The BLM, in cooperation with several state fish and game departments, employs this practice on important big-game ranges in Oregon, Utah, and Colorado.

Recreational and aesthetically oriented benefits are derived from the effects of livestock management. Certain areas of the southwestern desert are recognized for exceptional wildflower displays which require protection from grazing at certain seasons. In the same region, management is needed to prevent large concentrations of livestock and unsightly damage on ephemeral ranges. Improved access and use for the general public frequently results because of the construction of such facilities as roads and watering locations.

The present BLM range management program basically involves range-use administration, use adjustments, allotment management planning, range studies, and range supervision. Because the BLM is faced with the responsibility of proper resource use and improvement in deteriorated conditions, a primary long-term objective of the range management program is to control livestock use in order to improve range condition and thereby meet wildlife habitat and watershed objectives. This is accomplished by the application of sound scientific principles and intensive management through an allotment management plan.

GRAZING MANAGEMENT PLANNING

An allotment management plan (AMP) is a plan to meet specific needs and objectives. Preparation and implementation of an AMP is an on-the-ground action program. It involves identification of specific conservation and production objectives that can be accomplished by livestock grazing (identified in a multiple-use context), initiation of a grazing system designed to reach those objectives, and procedures for evaluating progress. Implementation requires investment in planning, installation of required facilities, follow-up studies, and supervision.

Allotment management plans provide a long-term solution to proper management of domestic livestock for protection and improvement of the range resource base. At the same time, livestock use can usually be retained at existing levels. From an economic development standpoint, AMPs mean more and better forage, increased meat production, and improved wildlife and watershed conditions and aesthetics; they will contribute materially to the well-being of the land, the users and their communities, and the general public. Stocking rates may be increased as vegetative quantity and quality improve. The added forage production can be used to satisfy the dependent operator's qualified demand, help meet increasing forage needs, and contribute to local or regional economies. It is believed that public grazing lands have the potential for a threefold increase in forage production under proper management.

SUMMARY

THE domestic livestock grazing on the millions of acres of public domain lands is of little importance from a regional and local community point of view, but these public lands are vital for economic stability and growth. In a broader sense, livestock grazing is important in maintaining other values and natural resources that exist on wild lands. Livestock in the future may be kept on the public lands, not for meat production but as a natural tool to produce the vegetative conditions needed for wildlife habitat, soil protection, and other values. Domestic livestock grazing, therefore, has a large role to play in properly managing the environment.

REFERENCES

1. Upchurch, M. L. 1963. Public grazing lands in the economy of the West. *In* Land and Water Use, pp. 83–98. Am. Assoc. Adv. Sci. Publ. 73.

16

WILDLIFE AND GAME MANAGEMENT ON GRASSLANDS

LAWRENCE V. COMPTON,
WADE H. HAMOR, *and*
OLAN W. DILLON, JR.

RANGE WILDLIFE, like all wildlife, is public property. Much of its habitat, however, is on privately owned land and is therefore private property. The rancher can market access to these habitats and thus to the animals, but he cannot market the animals themselves. Some ranchers refer to such enterprises as the "marketing of wildlife." While not technically correct, the term enjoys local use and understanding.

Wildlife habitats on ranches are products of the rancher's management. They may be there by accident or design, but they exist at the discretion of the owner. He may choose to improve them, preserve them, or eliminate them.

Access to these wildlife habitats for the purpose of hunting or fishing is at the discretion of the owner. He has the right to prohibit access, grant free access, or market access rights in accordance with the needs of his ranching operation or his personal desires. To meet future needs for fish and wildlife, it is estimated that at least 50 percent of the ranchers should intensively practice habitat management. To interest this great a number, an attractive incentive should be offered. That incentive might be a reasonable but profitable fee for access to the products of the rancher's management efforts to be paid by sportsmen using the improved resource.

Habitat management, regardless of where it is practiced, is expensive. But ranchers are becoming more interested as the demand for game and for hunting and fishing space increases. They find that sportsmen will pay for quality opportunities.

The rancher does much to ensure an abundance of wildlife when he follows proven grazing practices and installs water conserva-

tion and ersosion control measures. Some manage habitat by simply
reducing livestock competition for plants that supply food and cover.
Others plant food and cover areas and build ponds. Some must make
other arrangements to favor an abundance of wildlife.

Hunters and fishermen are not strangers to the rancher; he too
is one or both. The planned management of habitat and the leasing
of access rights to these people may, however, be new to him. The
rancher's decision to enter or avoid the wildlife venture reflects his
desire to meet and deal with people, the time he has to manage and
care for sportsmen using his land, the monetary return that access
leasing must yield if he is to be satisfied, the pressure that added wildlife numbers may exert on his rangelands and croplands from which
he derives his major income, and his liability to the sportsmen using
his land.

The average rancher, if interested at all, moves slowly into the
new venture. He blends it gradually into his normal operation. His
investment in time and money is minimal until he is well acquainted
with its operation and potential. The aggressive "go-for-broke" men
invest heavily and quickly. Some profit, some fail, in marketing access
to wildlife.

The wise manager seeks the best technical advice available to
help him plan the use of his soils, water, plant, and animal resources.
His local soil and water conservation district, with its many connec-
tions with federal and state technical agencies, is a good source of
information.

SCHEDULING WILDLIFE HARVESTS

THE planning process includes an inventory of the ranch's natural
resources. In addition to soils, water, and plants it should list the
kinds and numbers of game and nongame species that are desired or
are attractive to potential paying customers. This inventory can be
as detailed or as general as the rancher wishes to make it. From an
annual inventory he can determine the numbers of animals or pounds
of fish that can and should be harvested each year. Next, he should
determine the level of average hunter and fisherman success in his
community. From these figures he arrives at the number of hunter-
days and fisherman-days his ranch can supply and the number of
sportsmen for whom he can supply a quality experience.

For example, a ranch on which an annual inventory was made

in late summer shows a population of 50 deer, 10 antelope, and 300 grouse. The ranch has 4 ponds of 16 acres that have harvestable trout and are used by wild ducks during fall migration. The rancher determines that a third of the deer and antelope should be harvested as well as all the trout that can be caught and about 40 percent of the grouse. He feels he can open the ponds to duck hunting on four mornings per week.

Local hunting success, according to the rancher's information, is 1 deer per 6 days of hunting, 1 antelope per 4 hunter-days, and 1.5 grouse per hunter-day. Fishing success amounts to 2 pounds per fisherman-day. Knowing his harvestable surplus and the rate of harvest, he determines that he can supply 96 man-days of deer hunting, 12 man-days of antelope hunting, and 80 man-days of grouse hunting. With a 2-man blind at each of the 4 ponds for a 30-day season, he has 144 man-days of duck hunting. At a yield to anglers of 50 pounds of trout per acre, he has 400 man-days of fishing.

The rancher then needs to know the number of sportsmen required to harvest his game crop. He finds that sportsmen in his area annually spend 4 days hunting deer, 3 days hunting antelope, 5 days each at grouse and duck hunting, and 8 days fishing. With this information he can determine the number of people needed. He divides the number of available recreation days for each species by the number of man-days sportsmen spend annually pursuing each species and finds that he needs 24 deer hunters, 4 antelope hunters, 16 grouse hunters, 28 duck hunters (based on the length of the season and blind capacity), and 50 fishermen (Table 16.1).

If he leases all his hunting and fishing to 24 deer hunters, it is probable that they could use all the hunting days available for antelope and grouse as well as deer. They might also use their full quota

TABLE 16.1. *Planning the wildlife harvest*

Species	Unit	Game surplus	Average harvest per man-day	Man-days available	Man-days spent yearly by sportsmen	Number of sportsmen required
Deer	number	16	.16	96	4	24
Antelope	number	3	.25	12	3	4
Grouse	number	120	1.50	80	5	16
Ducks	number	288*	2.00	144	5	28
Trout	lb	800	2.00	400	8	50

* Based on length of effective season and blind capacity.

of time hunting ducks and fishing. Even so, this would leave 24 man-days of duck hunting and 208 man-days of fishing. Wanting a full harvest of his game, the rancher may insist on leasing to a group of 30–35 sportsmen. At this number there still would be a surplus of 15–20 fishing days. These could be used by the sportsmen's families.

MANAGEMENT OF THE WILDLIFE ENTERPRISE

Costs associated with the wildlife venture deserve early consideration in the planning process. To provide productive habitat and good hunting and fishing on the example ranch, the rancher finds that he must defer grazing on 200 acres of land and restock two ponds every year. The loss of net grazing income is charged against the wildlife enterprise as is the cost of pond renovation, added liability insurance, advertising, service to sportsmen, and similar expenses that he may encounter. He might list his expenses as shown in Table 16.2.

If he leases access rights at $800, his return equals that usually gained from grazing the land. In other words, he gains nothing in extra income from the wildlife enterprise. While the decision on charges rests with the individual rancher, his profit from the land largely diverted to wildlife use should be at least double that usually returned from full utilization through grazing, in this case 200 acres and $300. The 100 percent increase in income from converted acres is considered minimum to initially attract sufficient numbers of ranchers into the habitat management field.

The rancher of this example—with this knowledge of his harvestable surplus, costs, and the returns needed—can approach sportsmen with a package deal amounting to about 330 man-days of hunting and 400 man-days of fishing at a price of about $1100 for the season.

He may elect to lease access on a daily or weekly basis or lease access separately to each of his hunting and fishing opportunities. Experience of others indicates that he might make more money through either of the latter two approaches, but his management problems may multiply beyond his capacity to solve them.

TABLE 16.2.   *Expenses in wildlife habitat management*

Annual expenses:	
Average net loss of grazing income from 200 acres @ $1.50 per acre	$300
Renovate and restock two ponds (8 acres) @ $20 per acre	160
Miscellaneous (insurance, advertising, service to sportsmen, etc.)	340
	$800

The individual rancher will decide the best way to lease hunting and fishing rights on his land. If his ranch is within 50 miles of a sizable population center, the rancher might do well to offer access by the day; if farther away, the weekly or seasonal lease might be best. Leasing by the day or week exposes the rancher to more customers, which, if he likes people, may be best for him. If his business occupies most of his time or he desires only limited contacts with those who hunt and fish on his land, then the seasonal lease would be better.

SUMMARY

RANGELAND wildlife is largely under the control of ranchers, as is the access to it. To meet future needs for recreation hunting it is estimated that 50 percent of the ranchers should practice wildlife habitat management. A monetary incentive, paid by sportsmen as an access fee, probably is the best means to accomplish the wildlife objective. The potential for ranchers to improve wildlife habitat and obtain a reasonable profit appears promising.

17

POLLUTION CONTROL:
CONTRIBUTION OF GRASSLANDS

B. D. BLAKELY

THE CONSERVATION of natural resources, improving our environment, and pollution abatement have become public concerns in a relatively short time. Environment and pollution are frequently discussed in the news media and in day-to-day conversation. The word "pollutant" has now been applied to diseases, insects, noise, radioactive substances, alkalinity, pesticides, and many other things which most of us have not thought of in that light.

For a long time it has been known that grasses, legumes, and other forage plants can play an important role in conserving natural resources, controlling crop diseases and insects, and offsetting some of the problems created by toxic mineral salts in the soil. More recently their role has been given emphasis in improving and protecting the environment and in preventing pollution from sources such as sediment, windblown soil, municipal and farm wastes, and some toxic substances.

Suddenly, we are asked for solutions to problems for which we have few answers and are not sure of those we do have. Yet we have some answers for many conditions that are polluting the environment and, in time, research and experience will give us more.

SEDIMENT

SEDIMENT as a pollutant is by far the greatest in volume that reaches our streams, lakes, and harbors. It is estimated to be 700 times greater than the solids in sewage. About 4 billion tons of sediment reach the streams each year, of which 1 billion are carried to the oceans; 380 mil-

196

lion tons are dredged from harbors and waterways. Not only do we lose the soil but it carries plant food elements, especially phosphorus, and some pesticides that also pollute the streams and lakes.

About 287 million acres of cropland are losing less than 5 tons of soil per acre per year. This is about 66 percent of the total cropland in the United States. Forage crops are doing their share in keeping losses down to this level. In addition, 9.6 million acres of cropland have been converted to grass since 1966.

On the other side of the ledger, about 37 million acres are losing from 11 to 20 tons per acre per year, and 18 million acres are losing more than 20 tons of soil per acre per year. Most of the latter loss occurs on land that should be converted to permanent vegetation. It is safe to assume that with such high annual losses per acre, a high percentage of the sediment finds its way to streams, lakes, reservoirs, and harbors. Short of making new land near the mouth of a river or picking up the soil at the bottom of slopes or in road ditches and spreading it back on the field, sediment cannot be recycled as can some other pollutants.

Since 1952 the Soil Conservation Service (SCS) has been using the universal soil loss equation and the wind erosion equation as guides in planning conservation systems in the eastern two-thirds of the country. The SCS is also using the universal equation to predict soil losses from construction sites. These two equations are the result of extensive research studies by the USDA Agricultural Research Service, and they have given us a way to judge the various alternative conservation systems that will reduce erosion to a level that can be tolerated.

Not all the sediment comes from farms and ranches. Much of it comes from housing and industrial developments, road construction, gullies, and streambanks themselves. It is not unusual to lose 100 tons per acre in one rain from a construction site that is not adequately protected with grasses and mechanical measures. Soil losses from land protected with a good cover of grasses and legumes are measured in pounds rather than tons. Table 17.1 reflects the benefits of grasses in reducing soil and water losses (2).

The use of various systems of "no tillage," in which crop residues are used as ground cover and there is no plowing or cultivating, has been very effective in reducing sediment in several parts of the country. This method of producing corn, soybeans, and sorghums has gained farmer acceptance in a relatively short time. In the southern states, corn is planted in stands of grasses such as tall fescue or ryegrass that

TABLE 17.1. *Effects of crops and cropping systems on runoff and soil losses*

Cropping system	Runoff	Soil loss
	(%)	*(tons/acre/yr)*
Continuous corn	18.7	38.3
Three-year rotations:		
Corn	12.6	18.4
Oats	9.9	10.1
Hay	3.8	5.4
Average for rotation	8.8	11.3
Continuous alfalfa	2.2	0.1
Kentucky bluegrass pasture	1.2	0.03

Source: Browning et al., 1948 (2).

have been suppressed with herbicides. The corn is harvested for silage, and the grass makes a regrowth for soil protection and pasture. "No tillage" can also be used where cover and green manure crops can be grown to give soil protection the year around.

App reports that a well-managed cover or green manure crop is equivalent to a ton of fertilizer (2). In this day of high rates of nitrogen fertilizer on some crops, a cover crop could, especially on sandy soils, use the excess nitrogen that otherwise would leach below the root zone.

Grasses and legumes have the same beneficial effects for nonagricultural uses as they do for sloping cropland where they are the key to stabilizing critical erosion sites. The main difference is that production is not the major factor as is the case on cropland and grasslands.

STRIP MINES

ABOUT 2 million acres of surface strip-mined areas in the United States are in need of some kind of vegetation or reshaping for lakes and other recreational uses. These mines occur on all kinds of soils, from rocky to clay. They are usually low in fertility and range from highly acid to neutral. Runoff from these scarred lands pollutes streams with sediment, acids, and minerals that affect the ecology. It has been estimated that as much as 3 million tons of sulfuric acid annually pollute about 5000 miles of streams from surface-mined areas in 10 states.

Of the 2 million acres, about 30 percent could be reclaimed for pasture and range. Many areas already have been reclaimed for grasslands and are now an asset to farmer and community.

URBAN AND INDUSTRIAL WASTES

RESEARCHERS have stated that future reuse of sewage effluents will not be a question of economics but one of necessity. Many cities and industries are using forage plants to recycle municipal sludge and process wastes; some have been doing so for several years. The city of Melbourne, Australia, has been using grasses in this way since 1893. In 1969, the city's 26,809-acre farm handled 97 percent of all the municipal wastes. This amounts to about 100 million gallons of sludge per day, or 369 million gallons per year at a cost of 62 cents per person in the metropolitan area. The farm has as many as 19,000 head of beef cattle, 50,000 head of sheep, and 120 horses. During the 1968–69 fiscal year 5363 beef animals and 52,356 sheep were sold in addition to 523 bales of wool and crutchings.

Before the sludge is applied, organic materials and mineral salts are removed so that they will not pollute the runoff or groundwater. Three methods of purification are used, depending on the season and rate of flow: (1) land filtration during periods of high evaporation, (2) grass filtration during periods of low evaporation, and (3) lagooning during peak daily flows and wet weather.

During the grazing season, most of the sludge is used to irrigate 10,713 acres of pasture. Four inches of effluent is applied about every 18 days. This requires between 180 to 250 acres (depending on the soil) to filter 1 million gallons per day. When the vegetation is not grazed, only 48 acres are needed to handle 1 million gallons per day. Pastures sown more than 50 years ago are yielding more than 30 tons of green feed annually.

Perennial and Italian ryegrasses; orchardgrass; and white, alsike, red, and strawberry clover make up the pasture mixture. Italian ryegrass is used on the grass filtration areas. Table 17.2 reflects the soil improvement before and after irrigation with sludge.

TABLE 17.2. *The effect on soil fertility of irrigation with sludge*

Soil fertility constituents	Before irrigation	After irrigation	
		12 years	26 years
		(ppm)	
Nitrogen	1,260	2,620	5,000
Phosphoric acid	450	1,700	2,500
Potash	1,540	8,010	10,920
Lime	600	3,200	3,900
Chlorine	420	260	210

The Campbell Soup Company at Paris, Tex., uses grasses to filter pollutants from wastewater used in vegetable preparation and processing in much the same way as the city of Melbourne. The main difference is that sprinkler irrigation is used on the 500-acre disposal area and the forage is cut for hay rather than grazed. The land was smoothed to eliminate low areas where water could concentrate. Terraces were constructed, and underground mains were laid with sprinkler heads about 135 feet above each terrace. The area was then seeded to reed canarygrass.

As much as 3.6 million gallons of water are discharged daily, after the grease and large pieces of vegetable matter are removed. As water enters the irrigation system, it contains 550–900 ppm of biological oxygen demand (BOD). Studies indicate that this overland flow treatment can achieve 99 percent reduction in BOD, 90 percent reduction in nitrogen and phosphorus, and complete color removal (3). These reductions are well within the standards set by the Federal Water Quality Control Administration. The one precaution the investigators mention is the buildup of salts that might occur after years of use.

Free-choice feeding trials showed that mature beef cattle and calves preferred the hay from the disposal system over Hay Grazer (a hybrid millet), johnsongrass, and bermudagrass. After four hours all the reed canarygrass hay was eaten, whereas only about half the bermudagrass and johnsongrass was eaten and a lesser amount of the Hay Grazer.

It is well known that plants will take up more nutrients if there is an excess. The plant nutrients in a single harvest of reed canarygrass amounted to 8000 pounds of phosphorus and nearly 50,000 pounds of nitrogen. This is significant inasmuch as they otherwise would have been available for algae growth in streams.

The Flushing Meadows Project at Phoenix, Ariz., is designed to handle the effluent from an activated sludge plant serving the metropolitan area. The objective is to reclaim sewage water that can be used to recharge the groundwater. The present capacity is 70,000 acre-feet per year, and it is expected to increase to 300,000 acre-feet by the year 2000.

The project consists of six basins, 20 by 700 feet in size. Four basins were seeded to bermudagrass in 1968, one was left bare, and one was covered with two inches of sand and topped with four inches of 3/8-inch gravel. Inundations range from two days wet and three days dry to two weeks wet and ten days dry. The vegetated basins gave

TABLE 17.3.   *Composition of sewage effluent and well water after percolation of effluent through the soil profile*

Constituents in water	Sewage effluent	Well water
	(ppm)	
Biochemical oxygen demand	20	0.2
Chemical oxygen demand	55	15.0
Ammonium (N)	25	1.9
Nitrate (N)	0.1	$(1–9)^w$
		$(20–30)^d$
Nitrite (N)	trace	$0.4^w–0.1^d$
Organic nitrogen	2	0.3
Phosphorus (P)	20	5
Boron	0.5	0.5
Total salts	900	950
Coliform bacteria, per cc	10^6	$(0–5)^d$
		$(8–33)^w$
Fecal coliform bacteria, per cc	10^6	$(0–2)^d$
		$(2–13)^w$

Note:
10^6 = 10 to sixth power = 10,000,000
w = wet
d = dry

the best infiltration rate and the gravel basin the lowest. Water samples from sewage effluent and a 30-foot well on the site gave the analysis shown in Table 17.3.

Day and Tucker reported that winter barley irrigated with sewage effluent produced 11.14 tons of forage per acre, wheat produced 12.71 tons, and oats produced 10.93 tons (4). This is an increase of 112, 263, and 249 percent respectively compared with pump water with no fertilization. In no case did pump water with 100 pounds of nitrogen and 75 pounds of phosphate outyield the sewage effluent. Barley was more sensitive to detrimental effects of sewage than were wheat and oats. Day and Tucker also reported in a later study that the effluent tended to lower grain yield and malting quality of barley.

It has been estimated that livestock waste in the United States feedlots and barns amounts to 2 billion tons per year. Much of this waste, of necessity, must be recycled back through crops. Grasses and other feed crops can be used for this purpose. This is being done in several places where feedlot and chicken operations are common. However, indications are that dehydrated animal wastes can be processed and blended with corn or soybean meal and fed back to the animals. At the Beltsville Research Center, in feeding trials with sheep, as much as 85 percent of barn wastes has been used in the ration. Similar results were obtained in Ohio feeding trials.

Using animal wastes to produce more grass for beef production or growing other crops has been profitable. In northern Georgia, animal wastes have been used to help transform the bare red clay soils to grass, resulting in improved fertility, less erosion, and better land use. From a 320-acre feedlot operation comprising about 90,000 head of beef cattle at Greeley, Colorado, manure is trucked to 10,000 acres for corn production, which is cut for silage and fed to the cattle at the feedlot.

While there are advantages to recycling animal manure through grasses, indications are that not all grasses are suitable for heavy applications. A heavy application of chicken litter on tall fescue is one that may be hazardous to the grazing animal. It can cause symptoms similar to nitrate poisoning, grass tetany (hypomagnesia), and fat necrosis.

Nitrate poisoning is caused when nitrates are reduced to nitrite, which when combined with hemoglobin lowers the ability of blood to carry oxygen. The animal will pant continuously and will die unless treated with methylene blue. Hypomagnesia affects the nerves and muscles and is caused by high potassium and low magnesium. Both nitrate poisoning and hypomagnesia may also result if heavy applications of inorganic nitrogen fertilizers are used on small-grain pastures and grasses that are growing rapidly. Injecting magnesium into the bloodstream is a cure. Fat necrosis also may occur when animals graze pastures, over a prolonged period, that have been fertilized with 10 or more tons of chicken litter per acre. Masses of hard fat can accumulate in the abdominal cavity of the cow, which can interfere with calving and can cause strangulation of the intestines and eventually death.

Tall fescue, when fertilized with chicken litter, accumulates nitrates much more than does bermudagrass, and fat necrosis is likely to occur. To date there is no evidence that this fat necrosis will occur as the result of grazing bermudagrass. Limited research trials indicate that not more than 4 tons of chicken litter per acre per year should be used on fescue, whereas 20 tons can be used on bermudagrass. It is evident that much more research is needed to determine the effects of heavy applications of manure on different grasses. The question, Do all warm-season grasses react the same way as bermudagrass, and do all cool-season grasses react the same as tall fescue? may be raised. In literature dealing with recycling animal and municipal wastes through the soil and plants, authors frequently caution against overloading the system and emphasize the importance of continued trials on the fate of toxic substances in the soil and plants and in the grazing animals.

SUMMARY

THERE COULD very well be an upsurge in the use of forages for preventing pollution of the environment—soil, water, and air. Land-use regulations limiting the way soils may be used may become a reality. When this happens, vegetation will have an even greater role.

REFERENCES

1. App, Frank. 1956. The value of green manure crops in farm practice. *Better Crops with Plant Food* (April).

2. Browning, G. M., R. A. Norton, A. G. McCall, and F. G. Bell. 1948. Investigation in erosion control and the reclamation of eroded land . . . , Clarinda, Iowa, 1931–42. USDA Tech. Bull. 959.

3. Day, A. D., and T. C. Tucker. 1959. Production of small grain pasture forage using sewage effluent as a source of irrigation water and plant nutrients. *Agron. J.* 51:569–72.

4. Thornthwaite, C. W., et al. 1969. An evaluation of cannery waste disposal by overland flow spray irrigation, Campbell Soup Company, Paris [Texas] plant. *Climatology* 22 (2): 1–74.

18

GRASS FOR PROTECTION, SAFETY, BEAUTY, AND RECREATION

LLOYD E. PARTAIN

FEW WOULD DISAGREE that, of all plants, grass is the most important to man. That importance extends beyond economics. Grass is more than a direct or indirect source of food. The versatility of grass is unbounded. New uses and new forms are unending. Throughout the history of man, dependence on grass not only for sustenance and protection but for beauty, inspiration, leisure, and a symbol of well-being has prevailed. Its capability to serve multitudinous needs of mankind has been reflected in respect and reverence for it. Grass has been the stimulus for poetry and song; the solution to problems when all else fails; the attraction for exploration beyond the horizons of man's seemingly fixed environs; and it has served as the example for teaching the phenomena of perpetual regeneration, even the proof of immortality.

Perhaps no one better portrayed in descriptive terms man's dependence and appreciation for grass and its many and varied uses than did Senator John James Ingall of Kansas in 1872. Portions of his poetic prose, published in the *Kansas Farmer* magazine more than a century ago are cited here to set the tone for this chapter.

Next in importance to the divine profusion of water, light, and air, those three great physical facts which render existence possible, may be reckoned the universal beneficence of grass. Exaggerated by tropical heats and vapors to the gigantic cane congested with its saccharine secretion, or dwarfed by polar rigors to the fibrous hair of northern solitudes, embracing between these extremes the maize with its resolute pennons, the rice plant of southern swamps, the wheat, rye, barley, oats and other cereals, no less than the humbler verdure of hillside, pasture and prairie in the temperate zone, grass is the most widely distributed of all vegetable beings, and is at once the symbol of our life and the emblem of our mortality. Lying in the sunshine . . . of May, . . . our earliest recollections are of grass; and when the fitful fever is ended and the foolish wrangle of the market and forum is closed, grass heals over the scar which our descent into the bosom of the earth has made, and the carpet of the infant becomes the blanket of the dead.

What a beautiful way of saying grass is everywhere and exerts its powers and values upon us from the cradle to the grave.

Senator Ingalls related the references to grass as the theme in the teachings of great philosophers and as symbols of moralists. He quoted a prophet as saying, "All flesh is grass"; a troubled patriarch who sighed, "My days are as grass"; and reminded us that when in a penitential mood Nebuchadnezzar "did eat grass as oxen."

Being the keen observer and student that he was, Senator Ingalls then penned,

> Grass is the forgiveness of nature; her constant benediction. Fields trampled with battle, saturated with blood, torn with the ruts of cannon, grow green again with grass, and carnage is forgotten. Streets abandoned by traffic become grass-grown like rural lanes, and are obliterated. Forests decay, harvests perish, flowers vanish, but grass is immortal. Sown by the winds, by wandering birds, propagated by the subtle horticulture of the elements which are its ministers and servants, it softens the rude outline of the world. Its tenacious fibers hold the earth in place, and prevent its soluble components from washing into the wasting sea. It bears no blazonry or bloom to charm the senses with fragrance of splendor, but its homely hue is more enchanting than the lily or the rose. It yields no fruit in earth or air, and yet should its harvest fail for a single year, famine would depopulate the world.

The senator further sets the stage for discussion of values and challenges in the use of grass to improve the amenities of life for man when he says, "One grass differs from another grass in glory. There are grades of its vegetable nobility. Some varieties are useful. Some are beautiful. Others combine utility and ornament."

Among our efforts to alter the characteristics of life forms to fit our fancy (including genetic manipulation, mechanical controls, climatic adaptation, and a number of other rigorous and artificial forces) are those put to work in making grasses do what we want done for the enjoyment of people. For example, we use machines to rob a species or strain from its natural, sexually reproductive processes, and still it lives on in a form to suit our purpose. We transplant a variety or strain selection to regions beyond its indigenous environment where, without that natural support, we make it adapt to our whims. However, some grasses when grown beyond their natural ecological confines become problem children, or perhaps a better description would be "aggressive step-children." Where grasses become weeds, a wholly different research effort begins, that of control or utilization or both. Many of us have been prone to curse johnsongrass, bermudagrass, quackgrass, crabgrass, and other grasses when they have interfered with our primary goals for fields, golf courses, lawns, gardens, flower beds, and the like. Such dissatisfactions have challenged the agricultural chemists to produce selective herbicides to quiet our wailings and save our backs or our pocketbooks.

The matter of nourishing or combating grasses to make them perform to our liking is indeed big business. So are other commercial activities associated with the ever-broadening uses of grass for better living. This business is growing at such a rate that anyone attempting to apply a quantifying measurement upon it is likely to be frustrated because of lack of applicable data and information. Therefore this chapter does not deal in quantities for the most part, but it is hoped that some discussion of qualitative values may be worthwhile.

GRASS FOR SITE PROTECTION

MODERN MAN is continuously engaged in disturbing more and larger parcels of the earth to fit his complex ways of living. Protection of disturbed land space presents one of the great challenges of our time. For the most part, the resultant damage from such disturbance if left uncontrolled is in the form of erosion and sedimentation. Known methods for control and restoration of affected sites are largely vegetative and mechanical in principle.

Sediment produced by erosion is the most extensive pollutant of surface waters. It contributes heavily to pollution by suspended solids and intensifies the dissolved solids problem. Sediment is not only a pollutant in itself but it is a carrier of other harmful pollutants (chemicals, pathogens, and others) which impair the dissolved oxygen balance in water. Erosion that causes sediment is of nationwide concern.

The principal source of sediment is farm and forest lands where vast conservation programs are directed toward its reduction. But a large and increasing source of sediment is the nonfarm lands that are disturbed by man. Each year about one million acres are currently subject to erosion as they are in transition from farming and forestry uses to urban, transportation, industrial, and related develop-

mental uses. In addition to areas denuded by construction and development, thousands of miles of highways need erosion control both
during construction and in maintenance. The use of grass is prominent in such control.

Highway Construction. The use of grass in this nation's vast highway
        construction program has led to new specialization in the adapta-
        tion of knowledge about soils, hydrology engineering, and plant
science. It has brought increased attention to learning about the
properties of grass and other plant materials in preventing erosion
on a wide variety of site conditions. Two-thirds to three-fourths or
more of the finished right-of-way area is earthen surface. Much of it
if left bare is subject to severe erosion. Many such areas if left exposed become unsightly and dangerous. Maintenance costs soar. A
big part of the solution is in the power of grass, properly selected,
placed, and treated. Grass is a prime item in the development of a
modern highway system. This not only applies to new highway construction but establishment of grass is also needed along several hun-
dreds of thousands of miles of existing rural roads and secondary
highways for bank and cut stabilization and earth surfaces needing
protection.

    Even the most effectively engineered highway system creates off-
site problems. Wherever runoff is concentrated, severe erosion and
drainage problems may be created. The need for off-site grassed
waterways, water-spreading devices, and other techniques involving
the use of grass is extensive. Borrow areas are usually best restored
and made stable when adequately conditioned and planted to grass.
The proper use of grass contributes substantially to highway economy
by controlling erosion. Its adaptability to the efficient use of maintenance equipment and methods is an important factor. Grass helps
prevent the washing of soil from road cuts, slopes, and fills and the
sloughing due to frost action, thereby keeping sediment from clogging
culverts, drainageways, and other key structures in the system.

    Without stabilization with grass or other effective plantings,
erosion on many sites would eat into the roadbed itself or deposit
sediment on the highway surfaces. Such conditions are often aggravated by runoff and sediment coming onto highways from adjoining
lands. The solution to this problem is usually found in comprehensive planning in which highway designers and adjoining land-
owners work together to develop an erosion-control plan for a watershed or drainage area. Such cooperative endeavors are especially

adapted to overall planning and operation of small watershed protection and flood prevention projects established under Public Law
566, administered by the Soil Conservation Service of the USDA.
A grass-covered roadside or highway right-of-way protects adjoining farm or other open-land areas. Conversely, grassed areas off-site,
which drain onto highways, reduce erosion and sediment damage to
the highway system. Brant and Ferguson (1) put it succinctly when
they said, "A highway is more than a path for vehicles—it is a part
of the community, and roadside grass is a community asset in soil
conservation."

Grass will not solve all the erosion problems in disturbed areas
along highways. Great strides have been made, however, in both
highway location and design and in grass culture to extend effective
protection to difficult situations. The development of machinery and
equipment, fertilizers, mulching, and the like—along with the selection and improvement of various varieties and strains of grasses—
has recently focused on these difficult problem areas with substantial
success. Grass does not provide the panacea for all ills resulting from
disturbed earth. Some grades are too steep, some ditches and channels carry too much water at high velocities, and some shoulder areas
and other drive-off spots carry such congested traffic that even with
the best of grasses a sod cannot be established and maintained. Under
these conditions, man-made surfaces and structures must be used.

The difficulty of establishing grass under extremes of slope, soil
conditions, moisture availability, and other limiting factors is only
one aspect of controlling erosion and siltation on disturbed sites.
Grass, once established, needs continuous attention. In most areas,
regular fertilization is necessary to maintain an effective sod cover.
Mowing is usually needed to suppress competitive vegetation and to
maintain overall quality. Occasionally, some spots may need complete renovation including new seeding or sodding.

An ideal soil condition for the establishment of grass on disturbed sites is difficult to describe or achieve. The alternative, then,
is to learn the soil limitations site by site and apply the best information known about the grasses and grass culture. Soil and moisture
deficiencies must be offset by amendments and variety selection. One
of the best sources of information to guide the establishment of grass
cover on disturbed areas in a given locality is the local unit of the
Soil Conservation Service. Offices of such units, assigned to assist
local soil conservation districts, are located in about 3000 counties
in the United States.

Airfields. Much of that said about the use of grass in stabilizing disturbed areas along roads and highways applies to airfields. However, greater emphasis must be placed on the function of vegetation in checking dust and water erosion on airfields and airstrips.
Grass, strategically used, is usually found to be the most economical
and safe means for accomplishing this dual purpose. An installed
drainage system can be compatible with successful growing of grass
on such areas in most geographic regions. As more of the total land
area (runways, taxiways, parking aprons, service facilities, and access
roads) of an airport facility is surfaced with impervious materials,
drainage and vegetative practices on exposed soil portions become
more important and complicated. A tile drainage system is often
necessary. In very dry regions the collection and spreading of runoff
from surfaced areas can be advantageously utilized in grass culture,
provided damaging pollutants are held to tolerable levels.

Slopes or gradients for grassed areas on many soil types may be
an important factor to success. Soil, climate, and moisture manipulation will usually govern the varieties of grasses to be used as well
as the seeding rates. For the most part, heavier seeding rates than
those on agricultural lands are required. These are usually needed
to establish quickly and effectively the dense sod or turf needed for
heavy use and abuse, to resist wear, and to tolerate the pressure of
heavy loads. This is especially true in connection with airstrips.
Continued success with grasses in connection with airfields and airstrips depends on a high-level fertilization program, good maintenance, and in some instances supplementary irrigation.

Other Uses. The foregoing are merely examples of the use of grass
for protection of disturbed sites. The principles involved—with
due consideration to soil, water, and plant limitations—may be
applied to surface-mined areas, residential and industrial developments, and other nonfarm land uses.

GRASS AND SAFETY

Sports. The advantages and disadvantages of artificial grass over
natural turf for athletic fields has been widely discussed. Athletic officials, management and maintenance people, and specialists such as orthopedic surgeons and dermatologists are involved.
Opinions and feelings of participating athletes themselves have also

been expressed. Safety and comfort of the athlete seem to be the real issues. Regardless of the outcome of studies or the extent to which playing fields at super stadiums may convert to artificial grass, the outlook is for real grass footing for the majority of those engaging in football, baseball, golf, volley ball, and other sports historically using well-kept grass surfaces. Better field design, preparation, and maintenance along with better turf grasses for these purposes are still being sought.

Highways. Design, layout, and installation of many man-made structures consider the pairing of engineering and vegetation for safety's sake. Improvements in land shaping to meet this objective are constantly being made with the help of new concepts, new equipment, and new technology. A very high percentage of this shaped earth is being planted to grass.

Modern highways, for example, are designed with safety as an important factor in mind. Abruptness of incline, decline, curvature, and visual deterrents are avoided as much as possible. Construction of such design requires follow-up stabilization and maintenance. Surfaces must be kept intact if the designed conformation is to prevail over time. Grass helps accomplish the task. Rounded shoulders, well-sloped road banks and cuts, gullies and ravines filled and reshaped into gently rolling terrain, all well grassed, add to the safety of our highways. Grassed median strips and off-pavement side areas serve as refuge for vehicles troubled with malfunction or seeking to escape emergencies caused by others. Such areas have become a part of the basic plan for modern highways.

Judicious land shaping, plus effective seeding or sodding to well-adapted grass, can keep road shoulders and slopes from becoming rough and can prevent concentrated runoff from creating death-trap gullies. These and other grassed areas along highway rights-of-way and adjoining lands help to keep sediment and other erosional debris off the road and thereby prevent slippery conditions and other travel hazards.

There is increasing evidence that monotony of the landscape along highways, especially modern high-speed expressways, tends to create drowsiness for the driver, a hazard not only to himself but to others. To help overcome lapses into the drowsy feeling, the combination of grass, trees, and shrubs seems to have real value. Grass forms the base for interesting yet safe and attractive views for the motorist.

Soil surface protection, safety, and beauty are all a part of well-designed and well-executed travelways.

Recreation Areas. Grass is becoming more extensively used to help
   stabilize lakeshore and seaside areas where tides, rainstorms, and
   winds often damage or cause outright destruction to man-made
structures. Many "dream" cottages by the water's edge, boat docks,
improved beach areas, and other facilities for pleasant living become
little more than "bad dreams" because of the uncontrollable force of
water, wind, and shifting sands. Grass is no panacea for such
troubles, but with the use of some good sense in planning the location and construction of those facilities, grass can supply much added
protection. Selection of adaptable grass species, planting methods,
and protection of the grassed areas, once established, are factors
leading to success on these naturally difficult sites.

GRASS FOR BEAUTY

YEARS AGO, a Saturday chore in the South during most of the year
was to sweep the yard clean in both the front and rear of a farm
home. Packed earth kept barren by the use of the hoe and the
brush broom seems to have been a tradition. The shade of trees, the
heat of summer, and other limiting conditions perhaps precluded
the use of grass for a lawn. That tradition or practice has all but
disappeared as people have recognized that by use of adaptable
grasses, soil treatment, and fertilization a lawn becomes a rug on the
floor surrounding the home. The well-kept lawn becomes the cushion
for the total landscape—trees, shrubs, flowers, pathways and walks,
artistic fences, living screens, and windbreaks.

A well-kept lawn adds more than beauty to the landscape, it
brings satisfaction and pride to the family and increases the material
value of the home. To have a beautiful lawn, garden, trees, and
shrubs—all in a harmonious whole—requires thoughtful planning and
the expenditure of time, energy, and money. Often the natural environment must be altered. Some removal of dense shade may be
necessary. For most new homesites, especially in urban and suburban
areas, any semblance of natural soil conditions may have disappeared
as a result of excavation and construction. Tough clays or inert
sands may have to be completely reconditioned into simulated soil

for the successful establishment of a lawn. But once established, its nurture and upkeep can be a therapeutic diversion in the form of wholesome exercise; the profit is pride of accomplishment for better living.

Grasses for beauty need not be limited to lawns. Many species and strains are useful and attractive as individual bunches, clumps, or borders in the landscape. Some of the exotics are especially effective when used with certain types of architecture and garden designs. A good example is the use of specialty grasses to accent a Japanese or Spanish motif in the home and on the home grounds. More and more plant breeders and nurserymen are developing and offering grasses for a wide range of landscape uses. The growth in the number of turf farms and grass nurseries indicates this is indeed a booming business. Turf and grass-seed enterprises are to be found in practically all sections of the country. Associated with this trend has been the development of a wide range of varieties and strains tailored to variable growing conditions; the improvement in machinery and equipment, fertilizers, and irrigation; advances in weed-, disease-, and insect-control technology; and a growing appreciation for the beauty grass brings to the home, the community, and the landscape.

The beauties of grass are not limited to the homesite, the suburban neighborhood, parks, or playgrounds. Young (3) wrote, "Contour farming has long become a hallmark of soil conservation, a symbol of good husbandry, a creator of beauty, and a protector of the earth's great bounty." The curvaceous and patchwork pattern of farms in the gently rolling countryside forms one of rural America's more beautiful scenes. Grass in one form or another is grown in most of these contours. Young further stressed both the importance and beauty of this relatively new system of farming as he wrote, "Alternate strips of grain and grass following the contours of the landscape are subjects for the artist's canvas and at the same time protectors of the Nation's basic resource—protectors against soil erosion that robs the land of its productivity, pollutes our streams with mud, and shortens the life of our reservoirs."

Most people abhor the ugliness of streams and lakes laden with such pollutants as silt, sediment, sewage, and other wastes. Grass, used properly, can go a long way toward preventing or abating such pollution. Grass has been used for years to heal eroding hillside gullies, to prevent scouring of waterways and drainage systems, and to tie down soils subject to sheet erosion. We are learning more about grass-covered soil as filter fields to effectively dispose of sewage sludge, livestock manures, and certain processing wastes—all of which are harmful pollutants to our lakes and streams. More research and on-site technical help need to be devoted to this important aspect of pollution control. This will not only help beautify our lakes and streams but will also enhance the health and welfare of man.

A few years ago the governor of Vermont challenged agricultural, forestry, and recreation specialists to reverse the trend of so much land in his state reverting to tree cover. He pleaded for grassy, open fields and vistas to bring more tourists and vacationers, Vermont's second most important business, to that verdant country. In-depth studies are being conducted to find practical means for reserving open spaces, including pastoral scenes near large centers of population. Extensive areas of intermingled grasslands, trees, and other aesthetic land uses for the enjoyment of all are basic objectives.

No one appreciates the colors and forms of certain grasses in florescences, stems, and blades more than the flower arranger. For this artistic use we can add to the specialty grasses such domesticated species as wheat; maize, including colorful field corn; other cereals; and bamboo.

Inventories of seed-eating birds and the different grasses that provide an important source of their food would make long lists indeed. Moreover, grass provides nesting places and materials for a large number of birds and mammals that form an extensive part of the beauty and interest in the natural environment.

GRASS FOR RECREATION

THE SOIL SCIENTIST, the recreation planner, the golf architect, and other specialists engaged in site selection and development of outdoor recreation areas will generally have grass in the back of their minds when they choose the site for a playground, park, golf course, or campground. Even a cemetery or memorial park can be serenely beautiful. Grass has many values in or on recreation areas. First, of course, is attractiveness to which we quickly add protection of the site itself. Anyone who has had to pitch his tent or spread his bedroll on rough stony ground appreciates the comfort a soft, grassy spot offers by comparison.

The well-grassed playing field is ready for use soon after a rainstorm. One of the best uses to be made of a floodplain is for recreational activities that require a minimum of man-made structures, especially if the surface is maintained in permanent grass and trees. During dry weather a good grass cover will keep dust in parks and playgrounds to a minimum.

Perhaps no more grass specialism is used anywhere as in the design, construction, and operation of a good golf course. The fairway, the tee, the green, even the rough, each requires its own grass culture. So precise are the functions of different grassed areas on a good golf course that a very broad spectrum of interdisciplinary knowledge is required from start to finish. Moreover, the greenskeeper must be a rare combination of agrostologist, chemist, irrigation engineer, mechanic, pathologist, entomologist, and skilled artisan and also possess a number of other capabilities to keep the golfer happy. A plant that may not be a weed in any other place on earth may be absolutely forbidden on the green or even in the fairway. Nothing but grass, clipped at an absolutely precise height and rolled to the right degree of firmness, can be permitted to deter, hasten, obstruct, deflect, or guide that sensitive little sphere on its way to the cup.

Grass performs almost as complicated a job for the lawn tennis court. Even though most tennis is played on clay or other hard-surfaced courts, lawn tennis remains one of the most sophisticated of sports. The polo field is another sport facility requiring a well-sodded and maintained grass cover. Croquet, volley ball, badminton, and many other games are best enjoyed when played on grass. Bowling on the green, an ancient pastime, is still a popular sport in many parts of the world; interest in this skill game seems to be reviving in parts of the United States.

The following quotation is taken from a magazine featuring Hawaii (2): "On the south shore of Honaunau Bay, the king and his nobles lived, ruled and played. One can see the royal canoe launching sites and fishponds and even two Kahua Holuas. (The Kahua Holuas were like toboggan slides made of rocks covered with well-packed dirt and covered with slippery grass. This dangerous sport was reserved for royalty.)"

SUMMARY

THROUGHOUT the recorded history of man from the first gathering and storing of seeds for food, to the domestication of grass-eating animals, to the story of the shepherds and "Silent Night," to the ingenuity and enjoyment exemplified by the ancient Hawaiians, and

on down through the observant and eloquent recorders such as
Senator Ingalls, grass (oftentimes ruthlessly exploited) has provided
protection, safety, beauty, and pleasure to mankind. The story of
grass is a dynamic one. New uses and new forms of this most versatile
and important plant are yet to be found.

REFERENCES

1. Brant, Frank, and Marvin H. Ferguson. 1948. Safety and beauty for
highways. *In* Grass. USDA Yearbook Agr.
2. Northwest Orient Airlines. 1971. *Passages* (Sept.–Oct.).
3. Young, Gladwyn. 1967. Our land's golden splendor: Beauty and
bounty together. *In* Outdoors, U.S.A. USDA Handbook Agr.